ASCE STANDARD

American Society of Civil Engineers

Blast Protection of Buildings

This document uses both the International System of Units (SI) and customary units.

ASCE

AMERICAN SOCIETY OF CIVIL ENGINEERS

SEI

ASCE

**STRUCTURAL
ENGINEERING
INSTITUTE**

Published by the American Society of Civil Engineers

Cataloging-in-Publication data on file with Library of Congress

Published by American Society of Civil Engineers
1801 Alexander Bell Drive
Reston, Virginia 20191
www.pubs.asce.org

This standard was developed by a consensus standards development process which has been accredited by the American National Standards Institute (ANSI). Accreditation by ANSI, a voluntary accreditation body representing public and private sector standards development organizations in the U.S. and abroad, signifies that the standards development process used by ASCE has met the ANSI requirements for openness, balance, consensus, and due process.

While ASCE's process is designed to promote standards that reflect a fair and reasoned consensus among all interested participants, while preserving the public health, safety, and welfare that is paramount to its mission, it has not made an independent assessment of and does not warrant the accuracy, completeness, suitability, or utility of any information, apparatus, product, or process discussed herein. ASCE does not intend, nor should anyone interpret, ASCE's standards to replace the sound judgment of a competent professional, having knowledge and experience in the appropriate field(s) of practice, nor to substitute for the standard of care required of such professionals in interpreting and applying the contents of this standard.

ASCE has no authority to enforce compliance with its standards and does not undertake to certify products for compliance or to render any professional services to any person or entity.

ASCE disclaims any and all liability for any personal injury, property damage, financial loss or other damages of any nature whatsoever, including without limitation any direct, indirect, special, exemplary, or consequential damages, resulting from any person's use of, or reliance on, this standard. Any individual who relies on this standard assumes full responsibility for such use.

STANDARDS

In 2006, the Board of Direction approved the revision to the ASCE Rules for Standards Committees to govern the writing and maintenance of standards developed by the Society. All such standards are developed by a consensus standards process managed by the Society's Codes and Standards Committee (CSC). The consensus process includes balloting by a balanced standards committee made up of Society members and nonmembers, balloting by the membership of the Society as a whole, and balloting by the public. All standards are updated or reaffirmed by the same process at intervals not exceeding five years.

The following standards have been issued:

ANSI/ASCE 1-82 N-725 Guideline for Design and Analysis of Nuclear Safety Related Earth Structures

ASCE/EWRI 2-06 Measurement of Oxygen Transfer in Clean Water

ANSI/ASCE 3-91 Standard for the Structural Design of Composite Slabs and ANSI/ASCE 9-91 Standard Practice for the Construction and Inspection of Composite Slabs

ASCE 4-98 Seismic Analysis of Safety-Related Nuclear Structures

Building Code Requirements for Masonry Structures (ACI 530-02/ASCE 5-02/TMS 402-02) and Specifications for Masonry Structures (ACI 530.1-02/ASCE 6-02/TMS 602-02)

ASCE/SEI 7-10 Minimum Design Loads for Buildings and Other Structures

SEI/ASCE 8-02 Standard Specification for the Design of Cold-Formed Stainless Steel Structural Members

ANSI/ASCE 9-91 listed with ASCE 3-91

ASCE 10-97 Design of Latticed Steel Transmission Structures

SEI/ASCE 11-99 Guideline for Structural Condition Assessment of Existing Buildings

ASCE/EWRI 12-05 Guideline for the Design of Urban Subsurface Drainage

ASCE/EWRI 13-05 Standard Guidelines for Installation of Urban Subsurface Drainage

ASCE/EWRI 14-05 Standard Guidelines for Operation and Maintenance of Urban Subsurface Drainage

ASCE 15-98 Standard Practice for Direct Design of Buried Precast Concrete Pipe Using Standard Installations (SIDD)

ASCE 16-95 Standard for Load Resistance Factor Design (LRFD) of Engineered Wood Construction

ASCE 17-96 Air-Supported Structures

ASCE 18-96 Standard Guidelines for In-Process Oxygen Transfer Testing

ASCE/SEI 19-10 Structural Applications of Steel Cables for Buildings

ASCE 20-96 Standard Guidelines for the Design and Installation of Pile Foundations

ANSI/ASCE/T&DI 21-05 Automated People Mover Standards—Part 1

ANSI/ASCE/T&DI 21.2-08 Automated People Mover Standards—Part 2

ANSI/ASCE/T&DI 21.3-08 Automated People Mover Standards—Part 3

ANSI/ASCE/T&DI 21.4-08 Automated People Mover Standards—Part 4

SEI/ASCE 23-97 Specification for Structural Steel Beams with Web Openings

ASCE/SEI 24-05 Flood Resistant Design and Construction

ASCE/SEI 25-06 Earthquake-Actuated Automatic Gas Shutoff Devices

ASCE 26-97 Standard Practice for Design of Buried Precast Concrete Box Sections

ASCE 27-00 Standard Practice for Direct Design of Precast Concrete Pipe for Jacking in Trenchless Construction

ASCE 28-00 Standard Practice for Direct Design of Precast Concrete Box Sections for Jacking in Trenchless Construction

ASCE/SEI/SFPE 29-05 Standard Calculation Methods for Structural Fire Protection

SEI/ASCE 30-00 Guideline for Condition Assessment of the Building Envelope

SEI/ASCE 31-03 Seismic Evaluation of Existing Buildings

SEI/ASCE 32-01 Design and Construction of Frost-Protected Shallow Foundations

EWRI/ASCE 33-09 Comprehensive Transboundary International Water Quality Management Agreement

EWRI/ASCE 34-01 Standard Guidelines for Artificial Recharge of Ground Water

EWRI/ASCE 35-01 Guidelines for Quality Assurance of Installed Fine-Pore Aeration Equipment

CI/ASCE 36-01 Standard Construction Guidelines for Microtunneling

SEI/ASCE 37-02 Design Loads on Structures during Construction

CI/ASCE 38-02 Standard Guideline for the Collection and Depiction of Existing Subsurface Utility Data

EWRI/ASCE 39-03 Standard Practice for the Design and Operation of Hail Suppression Projects

ASCE/EWRI 40-03 Regulated Riparian Model Water Code

ASCE/SEI 41-06 Seismic Rehabilitation of Existing Buildings

ASCE/EWRI 42-04 Standard Practice for the Design and Operation of Precipitation Enhancement Projects

ASCE/SEI 43-05 Seismic Design Criteria for Structures, Systems, and Components in Nuclear Facilities

ASCE/EWRI 44-05 Standard Practice for the Design and Operation of Supercooled Fog Dispersal Projects

ASCE/EWRI 45-05 Standard Guidelines for the Design of Urban Stormwater Systems

ASCE/EWRI 46-05 Standard Guidelines for the Installation of Urban Stormwater Systems

ASCE/EWRI 47-05 Standard Guidelines for the Operation and Maintenance of Urban Stormwater Systems

ASCE/SEI 48-11 Design of Steel Transmission Pole Structures

ASCE/EWRI 50-08 Standard Guideline for Fitting Saturated Hydraulic Conductivity Using Probability Density Functions

ASCE/EWRI 51-08 Standard Guideline for Calculating the Effective Saturated Hydraulic Conductivity

ASCE/SEI 52-10 Design of Fiberglass-Reinforced Plastic (FRP) Stacks

ASCE/G-I 53-10 Compaction Grouting Consensus Guide

ASCE/EWRI 54-10 Standard Guideline for the Geostatistical Estimation and Block-Averaging of Homogeneous and Isotropic Saturated Hydraulic Conductivity

ASCE/SEI 55-10 Tensile Membrane Structures

ANSI/ASCE/EWRI 56-10 Guidelines for the Physical Security of Water Utilities

ANSI/ASCE/EWRI 57-10 Guidelines for the Physical Security of Wastewater/Stormwater Utilities

ASCE/T&DI/ICPI 58-10 Structural Design of Interlocking Concrete Pavement for Municipal Streets and Roadways

ASCE/SEI 59-11 Blast Protection of Buildings

CONTENTS

FOREWORD

The material presented in this publication has been prepared in accordance with recognized engineering principles. This Standard and Commentary should not be used without first securing competent advice with respect to their suitability for any given application. The publication of the material contained herein is not intended as a representation of warranty on the part of the American Society of Civil Engineers or of any person named herein, or that this information is suitable for any general or particular use or promises freedom from infringement of any patent or patents. Anyone making use of this information assumes all liability for such use.

The intent of the committee that prepared this standard was to present current practice in the analysis and design of structures for blast resistance. To accomplish that goal, the committee called upon its collective experience in the practice of blast resistant design, and consulted persons not on the committee. As such, this is a consensus document and does not reflect the specific practice of any individual.

This is the first edition of this standard. Its need had been identified in advance of the events of September 11, 2001. In fact, key individuals of the original nucleus of the committee and the Structural Engineering Institute (SEI) were on a conference call to discuss development of this standard as the events of that day began to unfold. In the months following SEI's announcement that a committee would be formed to prepare this standard, numerous experts stepped forward to volunteer for the effort. Hence, this standard represents approximately ten years of dedicated work by a knowledgeable committee.

The process started with subcommittees preparing "white papers" covering the information to be included in the standard. Once circulated for comment, those documents were reformatted into the first drafts of chapters of the mandatory and commentary sections of the standard. Then, throughout the development process the full committee reviewed and balloted numerous drafts of the standard. At each ballot cycle, subcommittees proposed resolutions for members' comments, ultimately leading to the full committee's approval of the text in this volume.

ACKNOWLEDGMENTS

Donald Dusenberry, P.E., F.ASCE, *Chair*
Jon Schmidt, P.E., M.ASCE, *Vice-Chair*
Paul Hobelmann, P.E., M.ASCE, *Secretary*

Chapter Leaders

Jon Schmidt, P.E., M.ASCE, Chapters 1, 2, and 3
Paul Mlakar, Ph.D., P.E., Dist.M.ASCE, Chapter 4
Lorraine Lin, Ph.D., P.E., M.ASCE, Chapters 5 and 8
Robert Smilowitz, Ph.D., P.E., M.ASCE, Chapters 6 and 7
Steven Smith, Ph.D., P.E., M.ASCE, Chapter 9
Andrew Whittaker, Ph.D., S.E., M.ASCE, Chapter 10

John Abruzzo, P.E., M.ASCE
Farid Alfawakhiri, Ph.D., P.E., M.ASCE
Iyad Alsamsam, Ph.D., P.E., S.E., M.ASCE
Charles Baldwin, P.E., M.ASCE
Curt Betts, P.E., M.ASCE
Scott Campbell, Ph.D., P.E., M.ASCE
Charles Carter, M.ASCE
Edward Conrath, P.E., M.ASCE
W. Corley, Ph.D., P.E., NAE,
    Dist.M.ASCE
Marvin Criswell, Ph.D., P.E., F.ASCE
Juan Carlos Esquivel, P.E., M.ASCE
Molly Evans, P.E., M.ASCE
David Fanella, Ph.D., P.E., F.ASCE
Simon Foo, P.E., M.ASCE

Andrew Hart, Aff.M.ASCE
Owen Hewitt, P.E., M.ASCE
Jennifer Holcomb, P.E., M.ASCE
Rolfe Jennings, P.E., M.ASCE
William Johnson, P.E., F.ASCE
Kim King, A.M.ASCE
Francis Laux, R.A., Aff.M.ASCE
Joel Leifer, P.E., M.ASCE
H. S. Lew, Ph.D., P.E., F.ASCE
Anatol Longinow, Ph.D., P.E., M.ASCE
Timothy Mays, A.M.ASCE
Douglas Merkle, Ph.D., P.E., F.ASCE
George Olive, M.ASCE
Glen Pappas, Ph.D., P.E., S.E., M.ASCE
Paul Perrin, P.E., M.ASCE

Keith Quick, P.E., M.ASCE
Ralph Rempel, P.E., M.ASCE
Hani Salim, A.M.ASCE
Karnik Seferian, P.E., M.ASCE
Joseph Smith, A.M.ASCE
Young Sohn, P.E.
Harold Sprague Jr., P.E., F.ASCE
Douglas Taylor
Andrew Thompson, P.E., M.ASCE
Gregory Varney, P.E., M.ASCE
Johnny Waclawczyk Jr., P.E., M.ASCE
Kenneth Walerius, P.E., M.ASCE
D. Erich Weerth, P.E., M.ASCE
William Zehrt Jr., P.E., M.ASCE
Mohamad Zineddin, M.ASCE

Chapter 1
GENERAL

1.1 SCOPE

This voluntary Standard provides minimum planning, design, construction, and assessment requirements for new and existing buildings subject to the effects of accidental or malicious explosions, including principles for establishing appropriate threat parameters, levels of protection, loadings, analysis methodologies, materials, detailing, and test procedures. However, this Standard is not applicable for the mitigation of potential accidents involving ammunition or explosives during their development, manufacturing, testing, production, transportation, handling, storage, maintenance, modification, inspection, demilitarization, or disposal.

This Standard is intended to supplement and not supersede the requirements of the governing building code and other applicable standards and laws. The omission of any specific material or system does not necessarily preclude its use in accordance with this Standard, as long as all applicable provisions are satisfied. This Standard does not prescribe requirements or guidelines for the mitigation of progressive collapse or other potential postblast behavior.

1.2 DEFINITIONS

The following definitions apply to the provisions of the entire Standard.

Aggressor: A person or organization that may initiate an attack against an asset.

Approved: Acceptable to the authority having jurisdiction.

Asset: A unit or collection of people or property that requires protection.

Attack: An attempt by an aggressor to cause the loss or compromise of an asset or group of assets.

Authority Having Jurisdiction: The organization, political subdivision, office, or individual charged with the responsibility of administering and enforcing the provisions of this Standard. It shall be permissible for the Authority Having Jurisdiction to be established by contractual agreement, when appropriate.

Average Strength Factor, ASF: A factor applied to nominal material strengths to account for the difference between the specified minimum and expected actual values. Also known as a **Static Increase Factor, SIF**.

Balanced Design: Controlled failure of a system with an established hierarchy of component failures, where connections are designed for the maximum strength of the connecting components and members supporting other members are designed for the maximum strength of the supported members. For window systems, the glazing shall fail before all other components.

Blast: Synonym for **Explosion**.

Building Envelope: Exposed elements on the exterior of the building, including (but not limited to) exterior walls, roofs, fenestration, exterior columns, spandrel and cantilever beams, and the exposed underside of occupied floors.

Buildings: Structures, usually enclosed by walls and a roof, constructed to provide support or shelter for an intended occupancy.

Compression Element: An element that carries an axial compression load greater than 10% of its axial compression strength. The factored load due to effects other than blast shall be determined in accordance with Section 3.5.3, and the effective strength shall be determined in accordance with Sections 3.5.1 and 3.5.2.

Connection: The means by which two or more elements are attached to each other, such as a beam to a column, a wall to another wall, a wall to a slab, etc. Steel connections are assemblies that include, but are not limited to, welds, bolts, rivets, angles, and plates. Reinforced concrete connections are often integral, consisting of the concrete and the reinforcement at the end of one element and extending into the other.

Consequence Factor: A numerical measure of the relative impact of the loss or compromise of a specific asset within a building, including its occupants, often expressed in terms of quantity or cost.

Constrained Fragment: A secondary fragment whose velocity in an airblast is reduced by the amount of energy required to tear it from its connected structural element.

Daylight Installation: A retrofit method for existing windows where security window film is applied to the interior vision surface of the glass without any additional attachment at the edges.

Dead Load, D: The weight of materials of construction incorporated into the building including, but not limited to, walls, floors, roofs, ceilings, stairways, built-in partitions, finishes, cladding, and other similarly incorporated architectural and structural items, and fixed service equipment including the weight of cranes.

Design Basis Threat: The explosive type and charge size for which the building is intended to provide a specified level of protection.

Diagonal Tension Shear: Shear associated with the flexural response of an element and the formation of diagonal cracks in reinforced concrete or masonry sections.

Direct Shear: Shear associated with the nearly instantaneous reaction force at the interface between connected elements in response to blast loading.

Ductile Flexural Element: An element that develops its plastic moment capacity and is capable of reliably sustaining deformation at or above this load level.

Ductility: A measure of the capability of an element, a cross section, or a connection to undergo inelastic deformation without significant loss of strength.

Ductility Ratio, μ: The ratio of the maximum deflection of an element to its maximum elastic deflection.

Dynamic Design Stress: The value used in place of the specified minimum yield stress to calculate the nominal strength of an element subject to blast loading, including any applicable strength increase factors, such as the average strength factor (ASF) and/or dynamic increase factor (DIF).

Dynamic Increase Factor, DIF: A factor, greater than 1, applied to expected material strengths to account for high strain rate effects.

Dynamic Response: Deformation, stress, and other behavior of a structure or structural element due to the action of a time-varying loading while considering inertial, stiffness, and, in some cases, damping effects.

Edge-to-Edge Installation: A retrofit method for existing windows where security window film applied to the interior vision surface of the glass extends into the window frame bite.

Elastic: Path-independent force-displacement behavior for loading and unloading, often idealized for structural materials by a linear force–displacement relationship, which results in no permanent deformation.

Elastoplastic: Path-dependent force-displacement behavior for loading and unloading, often idealized by a linear elastic branch up to the resistance at which first yield of the structural section occurs, successively reduced slopes as subsequent hinges develop, and a constant plastic resistance that is developed at the ultimate resistance of the section. The unloading path is usually assumed to follow an idealized linear elastic relationship.

Explicit Dynamic Finite Element Analysis: Finite element analysis in which dynamic response at a given time is expressed in terms of displacements, velocities, or accelerations at previous time-steps.

Explosion: A rapid chemical reaction that produces noise, heat, and violent expansion of gases.

Explosive: A material or device that is capable of causing an explosion under certain conditions, such as heat, shock, or friction.

Far Range: Standoff at which the blast loading from an explosion can be considered to be uniformly distributed over the tributary area of the element being loaded. This corresponds to a scaled distance Z that is equal to or greater than 3.0 $\text{ft/lb}^{1/3}$ (1.2 $\text{m/kg}^{1/3}$).

Flexural Element: An element that responds primarily in bending and does not carry an axial compression load greater than 10% of its axial compression strength. The factored load due to effects other than blast shall be determined in accordance with Section 3.5.3, and the effective strength shall be determined in accordance with Sections 3.5.1 and 3.5.2.

Frame Bite: A term used in glazing referring to the dimension by which the inner or outer edge of the frame or glazing stop overlaps the edge of the glazing.

Frangible: Attached to a protective structure but designed to fail during a blast event; typically lightweight and not posing a significant debris hazard. Common examples include portions of the structure itself, vent panels, canopies, exterior finishes, and exterior window shading devices.

Glazing Capacity: A calculated value of pressure and impulse at which glazing breakage initially occurs, which varies as a function of glazing dimensions, layup, and probability of glass failure.

Hazard-Based Design Approach: Design approach intended to provide exterior envelope components that reduce the risk of injury to the building occupants, recognizing that large portions of the exterior envelope may require replacement after an explosion.

Implicit Dynamic Finite Element Analysis: Finite element analysis in which the dynamic response of each node at a given time is expressed in terms of all the displacements, velocities, or accelerations at that time.

Impulse, _I_: Cumulative blast loading over time, calculated as the area beneath a pressure–time plot.

Joint: The region where two elements intersect, such as the region of the column to which a beam attaches. Steel joints consist of the panel zone and connections. Concrete joints consist of the volume that connects all intersecting columns and beams.

Level of Protection, LOP: The qualitative degree to which a building is expected to prevent or limit injury to and fatality of its occupants, and damage to and destruction of its contents, in the event of an explosion.

Live Load, _L_: A load produced by the use and occupancy of the building that does not include construction or environmental loads, such as wind load, snow load, rain load, earthquake load, flood load, or dead load.

Mechanical Attachment Systems: A retrofit method for existing windows where metal batten bars are attached to the window frame to retain the security window film after breakage of the window glass.

Mission: The purpose or function of the owner or users.

Near Range: Standoff at which the explosive is in close proximity to the structure relative to the size of the explosive, such that the resulting blast loading must be considered to be nonuniformly distributed over the tributary area of the element being loaded. This corresponds to a scaled distance Z that is less than 3.0 $\text{ft/lb}^{1/3}$ (1.2 $\text{m/kg}^{1/3}$).

Nonstandard Building Envelope Systems: Blast-mitigating systems for the building envelope that do not meet the definition of standard building envelope systems.

Nonstructural Elements: Non-load-bearing elements such as partitions, furniture, equipment, ceilings, and light fixtures.

Occupancy: The purpose for which a building, or part thereof, is used or intended to be used.

Owner: The person or organization that possesses or controls a particular asset, but does not necessarily directly utilize it on an ongoing basis.

Penetration: Disruption or displacement of some of the target material by a fragment that impacts but does not pass through the target.

Perforation: Passage through the target material by a fragment during impact.

Plastic: Path-dependent force-displacement behavior for loading and unloading associated with large deformations of structural materials, often idealized by a constant resistance function unless strain hardening of the materials is considered.

Primary Fragment: A fragment from a casing or container for an explosive source, or a fragment from an object in contact with an explosive.

Primary Structural Elements: The essential parts of the building's resistance to catastrophic blast loads and progressive collapse, including columns, girders, and the main lateral-force-resisting system, along with their connections, whose failure would likely result in disproportionate damage to or instability of the structure as a whole.

Progressive Collapse: Chain-reaction failure of a building's structural system or elements as a result of, and to an extent disproportionate to, initial localized damage, such as that caused by an explosion.

Recognized Literature: Published information including, but not limited to, research findings, technical papers, and reports that are approved by the Authority Having Jurisdiction.

Regular Shaped Structure: A structure having no unusual geometric irregularity in spatial form.

Resistance-Based Design Approach: Design approach intended to provide nonstructural exterior envelope components that fully resist specified blast loads.

Risk: The relative expected loss, accounting for consequence and vulnerability, due to a specific threat against a particular asset.

Scaled Distance, Z: The ratio of the standoff to the cube root of the explosive charge size as an equivalent mass of trinitrotoluene (TNT), which commonly serves as the basis for determining the blast loading parameters.

Secondary Fragment: A fragment produced by an object or structure located near the source of an explosion when the shock wave or a primary fragment encounters it.

Secondary Structural Elements: Load-bearing elements, including their connections, that are not primary structural elements.

Spalling: The formation of fragments on the side of an element facing away from the source of an explosion or fragment due to tension failure caused by the shock wave being transmitted through the element.

Standard Building Envelope Systems: Blast-mitigating systems for the building envelope with established design methodologies, analytical models with their performance correlated with blast testing, or systems that have been verified by full-scale blast testing in accordance with Chapter 10.

Standoff, R: The physical distance between the surface of a building, or part thereof, and the potential location for an explosion, such as the center of mass of an explosive charge.

Static Increase Factor, SIF: Alternative term for **Average Strength Factor, ASF.**

Story: The portion of a structure between the tops of two successive floor surfaces and, for the topmost story, from the top of the floor surface to the top of the roof surface.

Support Rotation, θ: The angle through which a flexural element subject to blast loading has rotated at its supports when it achieves its maximum dynamic deflection. When a flexural element is modeled for analysis as an equivalent single-degree-of-freedom (SDOF) dynamic system in accordance with this Standard, support rotation is calculated assuming straight segments between hinge or yield line locations and the point of maximum deflection.

Tactic: A method by which an aggressor could carry out an attack.

Threat Factor: A numerical measure of the relative likelihood of a particular explosive event.

Type I Cross-Section Flexural Resistance: The nominal moment capacity of a reinforced concrete or masonry element taking into account the compressive resistance of the concrete or masonry prior to crushing and including the strength increase effects of the compression reinforcement.

Type II and Type III Cross-Section Flexural Resistance: The nominal moment capacity of a reinforced concrete or masonry element after the concrete or masonry crushes. The resistance of the element is determined by the capacity of the reinforcing steel, neglecting the contribution of the concrete or masonry.

Unconstrained Fragment: A secondary fragment assumed to be loose or free to translate as a rigid body in an airblast.

Users: Persons or organizations that directly utilize a particular asset on an ongoing basis, but do not necessarily possess or control it.

Validated Constitutive Model: A constitutive model is validated for a range of calculations if it is known to consistently and repeatedly produce results and predictions in agreement with what has been observed in real situations covering that range of applications.

Vulnerability Factor: A numerical measure of the relative likelihood that a structural or nonstructural element will fail such that the loss or compromise of one or more assets, including people, would result. The vulnerability factor reflects the level of protection.

Wet-Glazed Installation: A retrofit method for existing windows where daylight or edge-to-edge film is used with a structural silicone bead attaching the security window film to the supporting frame.

1.3 SYMBOLS AND NOTATION

A = vent area of structure, ft^2 (m^2)

A_{ps} = area of prestressed reinforcement in tension zone, in.^{2} (mm^2)

ANFO = ammonium nitrate and fuel oil

ASF = average strength factor

b = width of compression face of reinforced or prestressed concrete element, in. (mm)

C_D = drag coefficient

C_r = sound velocity in reflected overpressure region, ft/msec (m/ms)

C_{ra} = reflection coefficient

CFD = computational fluid dynamics

COTS = commercial off-the-shelf

D = dead load determined in accordance with ASCE/SEI 7

d = effective depth of reinforced or prestressed concrete element, in. (mm)

DIF = dynamic increase factor

FRP = fiber reinforced polymer

f_{ps} = calculated stress in prestressing steel at design load, psi (MPa)

f'_c = specified compressive strength of concrete, psi (MPa)

H or h = building height, ft (m)

HFP = high-fidelity physics

i_g = gas overpressure impulse, psi-msec (MPa-ms)

i_r = normally reflected impulse, psi-msec (MPa-ms)

$i_{r\alpha}$ = reflected impulse for angle of incidence α, psi-msec (MPa-ms)

$i_{\bar{r}}$ = normally reflected negative impulse, psi-msec (MPa-ms)

i_s = incident impulse, psi-msec (MPa-ms)

L = live load determined in accordance with ASCE/SEI 7

LOP = level of protection

M_n = nominal flexural strength at section, lb-in. (N-m)

MDOF = multi-degree-of-freedom

P = overpressure, psi (MPa)

P_g = peak gas overpressure, psi (MPa)

P_r = peak normally reflected overpressure, psi (MPa)

$P_{r\alpha}$ = peak reflected overpressure for angle of incidence α, psi (MPa)

P_{so} = peak side-on or incident overpressure, psi (MPa)

$P_{\bar{r}}$ = peak normally reflected negative pressure, psi (MPa)

PETN = pentaerythritol tetranitrate, also known as penthrite

q_s = peak dynamic pressure, psi (MPa)

R = standoff, ft (m)

RDX = cyclotrimethylenetrinitramine, also known as cyclonite, hexogen, or T4

S = snow load determined in accordance with ASCE/SEI 7

SDOF = single-degree-of-freedom

t = time, msec (ms)

t_o = duration of positive phase, msec (ms)

t_{of} = equivalent triangular duration of positive phase incident pressure, msec (ms)

t_c = clearing time, msec (ms)

t_g = equivalent duration of gas overpressure, msec (ms)

t_{rf} = equivalent triangular duration of positive phase normally reflected pressure, msec (ms)

t_{rf}^{-} = equivalent triangular duration of negative phase normally reflected pressure, msec (ms)

TNT = trinitrotoluene

U_s = shock front velocity, ft/msec (m/ms)

V_f = interior free volume of structure, ft^3 (m^3)

V_u = direct or diagonal tension shear force, lb (N)

W = wind load determined in accordance with ASCE/SEI 7

W_e = explosive size as an equivalent quantity of TNT, lb (kg)

w = building width, ft (m)

Z = scaled distance, $R/W_e^{1/3}$, $ft/lb^{1/3}$ ($m/kg^{1/3}$)

α = angle of incidence between direction of propagation and loaded surface, degrees

γ_D = applicable dead load factor from Table 3-5

γ_L = applicable live load factor from Table 3-6

η = first-ply failure factor for FRP composite materials

θ = support rotation, degrees

θ_{max} = maximum permissible support rotation, degrees

μ = ductility ratio

μ_{max} = maximum permissible ductility ratio

ϕ = strength reduction factor

ω_p = reinforcement index for prestressed concrete element, $(A_{ps}/bd)(f_{ps}/f_c')$

$^\circ$ = degrees

1.4 QUALIFICATIONS

Evaluation of explosion effects requires specialized expertise in blast characterization, structural dynamics, and nonlinear behavior and numerical modeling of structures. This Standard is intended for use only by, or under the direct supervision of, licensed design professionals who are knowledgeable in the principles of structural dynamics and experienced with their proper application in predicting the response of elements and systems to the types of loadings that result from an explosion.

1.5 INFORMATION SENSITIVITY

Implementation of this Standard for a specific building is likely to result in the creation of information that could be useful to a potential aggressor in planning a malevolent attack. Such information shall be appropriately protected to the extent permissible by law.

1.6 CONSENSUS STANDARDS AND OTHER REFERENCED DOCUMENTS

The following references are consensus standards and are to be considered part of these provisions to the extent referred to in this chapter:

ASCE
American Society of Civil Engineers
1801 Alexander Bell Drive
Reston, VA 20191-4400
Minimum Design Loads for Buildings and Other Structures, ASCE/SEI 7, 2005.

Chapter 2
DESIGN CONSIDERATIONS

2.1 SCOPE

This chapter provides guidance for building owners and the users of this Standard to determine appropriate principles and design criteria for the mitigation of blast effects on buildings.

2.2 RISK ASSESSMENT

When not established by applicable law, owner policy, recognized industry standards, or other prescriptive means, structural design criteria for mitigating blast effects on a building shall be determined by means of a rational assessment of the risk of an explosion, whether accidental or malicious, in accordance with this chapter. The risk assessment shall be prepared by or in cooperation with qualified professionals in the fields of process safety and/or physical security as applicable, with the objective of identifying and prioritizing mitigation options.

2.2.1 Consequence Analysis. The risk assessment shall include identification and evaluation of the potential impacts of an explosion within or near the building, given the missions of the owner and users and the specific people and assets associated with the building that support these missions. The consequence factor shall be calculated or assigned to each such asset, including building occupants, on the basis of the effect of its loss on the owner's and users' missions; the time that would be required to replace the asset; and the relative value of the asset in terms of quantity, cost, or some other appropriate measure.

2.2.2 Threat Analysis. The risk assessment shall include identification of the potential causes of an explosion within or near the building and evaluation of the relative likelihood of each threat. Given the potential for collateral damage, the risk assessment shall take into account the presence of any neighboring facilities where an accidental or malicious explosion is considered likely to occur. It shall be permissible to establish multiple design basis threats for the building that are associated with different levels of protection as described in Section 3.3. In addition to airblast effects, the risk assessment shall determine whether it is appropriate to design the building or specific areas within it to mitigate the effects of secondary fragments in accordance with Chapter 5.

2.2.2.1 Accidental Threats. Each identified threat shall consist of a type, quantity, and location of explosives consistent with known or postulated sources of an accidental explosion. The threat factor shall be calculated or assigned on the basis of established scientific data or industry-specific mishap rates.

2.2.2.2 Malicious Threats. Each identified threat against a particular asset associated with the building shall consist of a known or postulated aggressor using a specific tactic and an associated type, quantity, and location of explosives consistent with the aggressor's anticipated capabilities. The threat factor shall be calculated or assigned on the basis of any relevant past incidents;

whether an attack on this asset would advance the aggressor's objectives; and whether intelligence data suggest that an attack on this asset is imminent.

2.2.3 Vulnerability Analysis. The risk assessment shall include identification of structural and nonstructural elements and operational procedures whose failure due to a specific threat would result in the loss or compromise of one or more assets associated with the building, and evaluation of the relative likelihood of such failure. The vulnerability factor shall be calculated or assigned on the basis of the level of protection selected for the relevant assets in accordance with Section 3.3.

2.2.4 Risk Analysis. The risk assessment shall include calculation or assignment of the relative risk for each combination of asset, threat, and building element, taking into account the calculated or assigned consequence, threat, and vulnerability factors. Protection of assets shall be prioritized with consideration of the ranked resulting values.

2.3 RISK REDUCTION

Mitigation measures other than structural design to reduce the building's risk due to blast loads shall be considered.

2.3.1 Consequence Reduction. The following measures shall be considered for the purpose of reducing the potential consequence of an accidental or malicious explosion that equals or exceeds the threats used for design: having multiple assets that can perform the same function; distributing assets among multiple facilities at different locations; providing an interior layout that minimizes the exposure of people and critical assets; adequately designing and supporting nonstructural components and systems; and contingency planning to facilitate incident response and recovery and continued operations.

2.3.1.1 Nonstructural Components and Systems. Design and support of nonstructural components and systems shall be consistent with the level of protection defined in accordance with Section 3.3. Components and systems that are required to remain operational after a blast event shall be designed and supported to withstand the design basis threat. The performance of such components and systems shall be qualified by either analysis and design or full-scale testing in accordance with Section 10.7.

2.3.2 Threat Reduction

2.3.2.1 Accidental Threats. The following measures shall be considered for the purpose of reducing the potential threat of an accidental explosion: remote storage of potentially explosive materials, and appropriate policies and procedures for safe handling of such materials.

2.3.2.2 Malicious Threats. The following measures shall be considered for the purpose of reducing the potential threat of a malicious explosion: standoff enforced by appropriate perimeter barriers; access control and searches of vehicles and persons approaching or entering the building; and enclosure of load-bearing primary structural elements in public areas to preclude direct contact with an explosive device.

2.3.2.2.1 Standoff. Where required by the risk assessment performed in accordance with Section 2.2, the standoff shall be maximized by means of site access controls combined with barriers such as anti-ram structures, site amenities, and landforms. Alternatively, it shall be permissible to use blast walls to reduce the effects of an explosion that occurs within or at the perimeter of the site. The performance of a blast wall, with respect to both airblast and fragmentation as applicable, shall be qualified by either analysis and design or full-scale testing in accordance with Section 10.5.

2.3.2.2.2 Vehicle Barriers. All anti-ram structures and site amenities intended to serve as vehicle barriers shall be designed to resist the vehicle threat identified by the risk assessment performed in accordance with Section 2.2. The maximum clear distance between an active anti-ram structure and the adjacent passive barriers, or between two adjacent passive barriers, shall be 4 ft (1.2 m). The minimum effective height of any vehicle barrier shall be 3 ft (0.9 m). The performance of a vehicle barrier shall be qualified by either analysis and design or full-scale testing in accordance with Section 10.3.

2.3.2.2.3 Landforms. It shall be permissible to use landforms as vehicle barriers, provided that they can stop the forward progress of the vehicle threat identified by the risk assessment performed in accordance with Section 2.2. The performance of a landform as a vehicle barrier shall be qualified by either analysis and design or full-scale testing in accordance with Section 10.3. Examples include ditches, berms, and bodies of water, which shall be permanently identified and properly maintained over time. However, such landforms shall not be arranged in a manner that would facilitate concealment of weapons or personnel within the site or otherwise degrade the effectiveness of other protective measures.

2.4 RISK ACCEPTANCE

This Standard is not intended to eliminate all risk associated with an accidental or malicious explosion. The building owner shall establish the level of residual risk that is acceptable based on a specific set of threats for which the building must be designed, the available budget for construction or renovation, or a combination of these considerations, consistent with the level of protection defined in accordance with Section 3.3.

Chapter 3
PERFORMANCE CRITERIA

3.1 SCOPE

This chapter provides the objectives of blast-resistant design and levels of protection that are associated with qualitative descriptions of damage to structural and nonstructural elements and corresponding response limits. Analysis and design requirements are specified in Chapters 6 through 8. Detailing requirements are specified in Chapter 9.

3.2 DESIGN OBJECTIVES

The primary purpose of blast-resistant design is to reduce, to a defined extent, the risk to occupants of injury and fatality and to contents of damage and destruction in the event of an explosion of a specified magnitude and location within or near the building.

3.2.1 Limit Structural Collapse. All structural elements shall be designed and detailed to respond in a manner consistent with the defined level of protection to the direct and indirect effects of the specified explosive threats in accordance with this Standard. When these blast effects are expected to cause plastic hinging or localized failure of individual structural elements, the damaged state of the structural system as a whole shall be evaluated to verify that global stability is maintained.

3.2.2 Maintain Building Envelope. All exterior structural and nonstructural elements, including openings, shall be designed and detailed to reduce the potential of a breach that would allow the overpressures from the specified exterior explosive threats to enter the interior of the building, consistent with the defined level of protection. For façade components, including windows and doors, both resistance-based and hazard-based design approaches shall be acceptable.

3.2.3 Minimize Flying Debris. Barriers, site furnishings, landscaping features, and structural and nonstructural elements, including exterior openings such as windows and doors, and interior overhead mounted items, shall be located, designed, and detailed to reduce the potential for producing hazardous secondary fragments due to the specified explosive threats, consistent with the defined level of protection.

3.3 LEVELS OF PROTECTION

The required level of protection (LOP) shall be defined in accordance with Sections 2.2, 2.3, and 2.4 for the building as a whole or each portion thereof and for each specific element, taking into account use and occupancy considerations, consistent with the following performance goals:

- LOP I (Very Low): Collapse prevention; surviving occupants will likely be able to evacuate but the building is unlikely to be safe enough for them to return; contents may not remain intact.

- LOP II (Low): Life safety; surviving occupants will likely be able to evacuate and then return only temporarily; contents will likely remain intact for retrieval.

- LOP III (Medium): Property preservation; surviving occupants may have to evacuate temporarily but will likely be able to return after cleanup and repairs to resume operations; contents will likely remain at least partially functional but may be impaired for a time.

- LOP IV (High): Continuous occupancy; all occupants will likely be able to stay and maintain operations without interruption; contents will likely remain fully functional.

3.3.1 Structural Damage. The potential overall structural damage for each level of protection shall be as follows:

- LOP I (Very Low): Damage is expected, up to the onset of total collapse, but progressive collapse is unlikely.

- LOP II (Low): Damage is expected, such that the building is not likely to be economically repairable, but progressive collapse is unlikely.

- LOP III (Medium): Damage is expected, such that the building is likely to be economically repairable, and progressive collapse is unlikely.

- LOP IV (High): Only superficial damage is expected.

3.3.2 Element Damage. The potential damage to individual elements for each LOP shall be as follows and as indicated in Table 3-1:

- Hazardous: The element is likely to fail and produce debris.
- Heavy: The element is unlikely to fail but will likely have significant permanent deflections such that it is unlikely to be repairable.
- Moderate: The element is unlikely to fail but will likely have some permanent deflection such that it is likely to be repairable, although replacement may be preferable for economic or aesthetic reasons.
- Superficial: The element is unlikely to exhibit any permanent deflection or visible damage.

3.3.3 Glazing Behavior. The potential glazing hazard for each LOP shall be as follows:

- LOP I (Very Low): Low Hazard in accordance with ASTM F1642.
- LOP II (Low): Very Low Hazard in accordance with ASTM F1642.
- LOP III (Medium): Minimal Hazard in accordance with ASTM F1642.
- LOP IV (High): No Break in accordance with ASTM F1642.

3.3.4 Door Behavior. The potential door hazard for each LOP shall be as follows:

- LOP I (Very Low): Doors are likely to be dislodged from their frames.

TABLE 3-1. EXPECTED ELEMENT DAMAGE FOR EACH LEVEL OF PROTECTION

Level of Protection	Primary Structural Elements	Secondary Structural Elements	Nonstructural Elements
I (Very Low)	Heavy	Hazardous	Hazardous
II (Low)	Moderate	Heavy	Heavy
III (Medium)	Superficial	Moderate	Moderate
IV (High)	Superficial	Superficial	Superficial

- LOP II (Low): Category IV in accordance with ASTM F2247. Doors are likely to be inoperable and wedged in their frames.
- LOP III (Medium): Category III in accordance with ASTM F2247. Doors are likely to be operable only by removing hardware.
- LOP IV (High): Category I or II in accordance with ASTM F2247. Doors are likely to remain operable but may require moderate force.

3.4 RESPONSE LIMITS

For analysis of individual elements in accordance with this Standard, appropriate response limits for all applicable modes of failure shall be established using a rational procedure that accounts for the damage or hazard associated with each LOP as described in Section 3.3.

3.4.1 Flexural Elements. When a flexural element subject to far-range blast effects is modeled for analysis as an SDOF dynamic system in accordance with this Standard, the ductility ratio or support rotation used to determine the LOP provided against a specified explosive threat shall not exceed the applicable value provided in Table 3-2. Where both values are provided, both limits shall be satisfied. Before applying these limits, the analysis shall demonstrate design adequacy for all other potential failure modes, including flexural and direct shear, or as a result of load reversal or element rebound.

3.4.2 Compression Elements. When a compression element subject to far-range blast effects is modeled for analysis as an SDOF dynamic system in accordance with this Standard, the ductility ratio or support rotation used to determine the LOP provided against a specified explosive threat shall not exceed the applicable value provided in Table 3-3. Where both values are provided, both limits shall be satisfied. Before applying these limits, the analysis shall demonstrate design adequacy for all other potential failure modes, including flexural and direct shear, or as a result of load reversal or element rebound.

3.5 ELEMENT STRENGTH

For analysis of individual elements in accordance with this Standard, the element resistance function shall include appropriate strength increase factors, strength reduction factors, and effects other than blast.

3.5.1 Strength Increase Factors. It shall be permissible to determine the nominal strength of an element using a rational procedure that accounts for the expected actual material properties of the element, rather than the specified minimum properties; the estimated strain rates in the element due to blast effects; and other relevant aspects of the dynamic element response. When an element is modeled for analysis as an SDOF dynamic system in accordance with this Standard, it shall be permissible to mul-

tiply each material property that contributes to the strength of the element by the following factors to determine the nominal strength of the element:

- The applicable average strength factor (ASF) provided in Table 3-4.
- The applicable dynamic increase factor (DIF) provided in Table 3-5.

3.5.2 Strength Reduction Factors. The nominal strength of an element shall be calculated in accordance with load and resistance factor design principles and multiplied by a strength reduction factor ϕ to determine the design strength of the element. For all modes of failure, it shall be permissible to use a strength reduction factor ϕ equal to 1.0.

3.5.3 Remaining Strength. The design strength of an element shall be reduced by the effects of the following load combinations using appropriate material-specific interaction equations in order to determine the remaining strength that is available to resist blast loads:

$$\gamma_D D + (\gamma_L L \text{ or } 0.2S) \tag{3-1}$$

$$\gamma_D D + 0.2W \tag{3-2}$$

where

γ_D = applicable dead load factor from Table 3-6
γ_L = applicable live load factor from Table 3-7
D = dead load determined in accordance with ASCE/SEI 7
L = live load determined in accordance with ASCE/SEI 7
S = snow load determined in accordance with ASCE/SEI 7
W = wind load determined in accordance with ASCE/SEI 7.

Equation 3-2 is not applicable when the stability of the structure is checked by imposing a notional lateral force equal to or greater than 0.2% of the total gravity force due to the summation of the dead and live loads acting on the story above that level. In no case shall the effective mass of a dynamically responding element be taken greater than the actual mass that provides inertial resistance to deformation of the element.

3.6 CONSENSUS STANDARDS AND OTHER REFERENCED DOCUMENTS

The following references are consensus standards and are to be considered part of these provisions to the extent referred to in this chapter:

ACI
American Concrete Institute
38800 Country Club Drive
Farmington Hills, MI 48331
Building Code Requirements for Structural Concrete, ACI 318, 2008.

AISC
American Institute of Steel Construction
One East Wacker Drive, Suite 700
Chicago, IL 60601-1802
Specification for Structural Steel Buildings, ANSI/AISC 360, 2010.

ASCE
American Society of Civil Engineers
1801 Alexander Bell Drive
Reston, VA 20191-4400
Minimum Design Loads for Buildings and Other Structures, ASCE/SEI 7, 2005.

TABLE 3-2. MAXIMUM RESPONSE LIMITS FOR SDOF ANALYSIS OF FLEXURAL ELEMENTS[a]

Element Type	Superficial		Moderate		Heavy		Hazardous	
	μ_{max}	θ_{max}	μ_{max}	θ_{max}	μ_{max}	θ_{max}	μ_{max}	θ_{max}
Reinforced Concrete								
Single-reinforced slab or beam[b]	1	–	–	2°	–	5°	–	10°
Double-reinforced slab or beam without shear reinforcement[c,d]	1	–	–	2°	–	5°	–	10°
Double-reinforced slab or beam with shear reinforcement[c]	1	–	–	4°	–	6°	–	10°
Prestressed Concrete[d]								
Slab or beam with $\omega_p > 0.30$	0.7	–	0.8	–	0.9	–	1	–
Slab or beam with $0.15 \le \omega_p \le 0.30$	0.8	–	$0.25/\omega_p$	1°	$0.29/\omega_p$	1.5°	$0.33/\omega_p$	2°
Slab or beam with $\omega_p \le 0.15$ and without shear reinforcement[c,d]	0.8	–	$0.25/\omega_p$	1°	$0.29/\omega_p$	1.5°	$0.33/\omega_p$	2°
Slab or beam with $\omega_p < 0.15$ and Shear reinforcement[c]	1	–	–	1°	–	2°	–	3°
Masonry								
Unreinforced[d,f]	1	–	–	1.5°	–	4°	–	8°
Reinforced	1	–	–	2°	–	8°	–	15°
Structural Steel (Hot-Rolled)								
Beam with compact section[g]	1	–	3	3°	12	10°	25	20°
Beam with noncompact section[d,g]	0.7	–	0.85	3°	1	–	1.2	–
Plate bent about weak axis	4	1°	8	2°	20	6°	40	12°
Open Web Steel Joist								
Downward loading[h]	1	–	–	3°	–	6°	–	10°
Upward loading[i]	1	–	1.5	–	2	–	3	–
Shear response[j]	0.7	–	0.8	–	0.9	–	1	–
Cold-Formed Steel								
Girt or purlin	1	–	–	3°	–	10°	–	20°
Stud with sliding connection at top	0.5	–	0.8	–	0.9	–	1	–
Stud connected at top and bottom[k]	0.5	–	1	–	2	–	3	–
Stud with tension membrane[l]	0.5	–	1	0.5°	2	2°	5	5°
Corrugated panel (1-way) with full tension membrane[m]	1	–	3	3°	6	6°	10	12°
Corrugated panel (1-way) with some tension membrane[n]	1	–	–	1°	–	4°	–	8°
Corrugated panel (1-way) with limited tension membrane[o]	1	–	1.8	1.3°	3	2°	6	4°
Wood[p]	1	–	2	–	3	–	4	–
FRP Composites	Response limits shall be determined in accordance with Section 9.6							
Glazing System Framing								
Aluminum	1	–	5	3°	7	6°	10	10°
Steel	1	–	–	3°	–	6°	–	10°

[a]Where a dash (–) is shown, the corresponding parameter is not applicable as a response limit.
[b]Reinforcement ratio shall not exceed 0.5 times the value that produces balanced strain conditions, as defined by section 10.3.2 of ACI 318.
[c]Stirrups or ties that satisfy Sections 11.4.5 and 11.4.6 of ACI 318 and enclose both layers of flexural reinforcement throughout the span length.
[d]These response limits are applicable for flexural evaluation of existing elements that satisfy the design requirements of Chapters 6 through 8 but do not satisfy the detailing requirements in Chapter 9, and shall not be used for design of new elements.
[e]Reinforcement index $\omega_p = (A_{ps}/bd)(f_{ps}/f'_c)$
[f]Values assume wall resistance controlled by brittle flexural response or axial load arching with no plastic deformation; for load-bearing walls, use Superficial or Moderate damage limits to preclude collapse.
[g]Limiting width-to-thickness ratios for compact and noncompact sections are defined in ANSI/AISC 360.
[h]Values assume tension yielding of bottom chord with adequate bracing of top chord to prevent lateral buckling.
[i]Values assume adequate anchorage to prevent pull-out failure and adequate bracing of bottom chord to prevent lateral buckling.
[j]Applicable when element capacity is controlled by web members, web connections, or support connections; ductility ratio for shear is equal to peak shear force divided by shear capacity.
[k]Also applicable when studs are continuous across a support.
[l]Requires structural plate-and-angle bolted connections at top, bottom, and any intermediate supports.
[m]Sheet has adequate connections to yield cross section fully.
[n]Typically applicable for simple-fixed-span conditions.
[o]Limited to connector capacity; includes all standing seam metal roof systems.
[p]Values shown are based on very limited testing data; use specific test data if available.
SDOF, single degree of freedom; FRP, fiber reinforced polymer; μ, ductility ratio; θ, support rotation.

Element Type	Superficial		Moderate		Heavy		Hazardous	
	μ_{max}	θ_{max}	μ_{max}	θ_{max}	μ_{max}	θ_{max}	μ_{max}	θ_{max}
Reinforced Concrete								
Single-reinforced beam-column[b]	1	–	–	2°	–	2°	–	2°
Double-reinforced beam-column without shear reinforcement[c,d]	1	–	–	2°	–	2°	–	2°
Double-reinforced beam-column with shear reinforcement[c]	1	–	–	4°	–	4°	–	4°
Wall or seismic column[e,f]	0.9	–	1	–	2	–	3	–
Nonseismic column[e,f]	0.7	–	0.8	–	0.9	–	1	–
Masonry								
Unreinforced[d,g]	1	–	–	1.5°	–	1.5°	–	1.5°
Reinforced	1	–	–	2°	–	2°	–	2°
Structural Steel (Hot-Rolled)								
Beam-column with compact section[h,i]	1	–	3	3°	3	3°	3	3°
Beam-column with noncompact section[h,i]	0.7	–	0.85	3°	0.85	3°	0.85	3°
Column (axial failure)[f]	0.9	–	1.3	–	2	–	3	–
Wood[j]								
Beam-column (flexural failure)	1	–	2	–	2	–	2	–
Column (axial failure)[f]	–	–	–	–	–	–	1	2.4°
FRP Composites	Response limits shall be determined in accordance with Section 9.6							

[a]Where a dash (–) is shown, the corresponding parameter is not applicable as a response limit.
[b]Reinforcement ratio shall not exceed 0.5 times the value that produces balanced strain conditions, as defined by section 10.3.2 of ACI 318.
[c]Stirrups or ties that satisfy Sections 11.4.5 and 11.4.6 of ACI 318 and enclose both layers of flexural reinforcement throughout the span length.
[d]These response limits are applicable for evaluation of existing elements only and shall not be used for design of new elements.
[e]Seismic columns have ties or spirals that satisfy, at a minimum, the requirements of Section 21.3.5 of ACI 318; see Chapter 9 for complete detailing requirements.
[f]Ductility ratio is based on axial deformation, rather than flexural deformation.
[g]Values assume wall resistance controlled by brittle flexural response or axial load arching with no plastic deformation; for load-bearing walls, use Superficial or Moderate damage limits and restrict deflection to 1/6 of the wall thickness to preclude collapse.
[h]Limiting width-to-thickness ratios for compact and noncompact sections are defined in ANSI/AISC 360 (ANSI 2005).
[i]If a shear plane through the anchor bolts connecting the column base plate to the foundation exists, the response limit for superficial damage shall apply, using the shear capacity of this connection, rather than the element flexural capacity, as the ultimate resistance for analysis.
[j]Values shown are based on very limited testing data; use specific test data if available.
SDOF, single degree of freedom; FRP, fiber reinforced polymer; μ, ductility ratio; θ, support rotation.

TABLE 3-4. AVERAGE STRENGTH FACTORS (ASF) FOR SDOF ANALYSIS

Material and Property	ASF
Compressive strength of concrete	1.1
Yield strength of reinforcing steel, hot-rolled steel ≤50 ksi (350 MPa), or cold-formed steel	1.1
All other materials and properties, unless another value is established on the basis of actual data	1.0

SDOF, single degree of freedom.

TABLE 3-5. DYNAMIC INCREASE FACTORS (DIF) FOR SDOF ANALYSIS[a]

Material and Property	Failure Mode	DIF
Concrete compressive strength	Flexure	1.19
	Compression	1.12
	Direct Shear	1.10
Masonry compressive strength	Flexure	1.19
	Compression	1.12
	Direct Shear	1.10
Deformed reinforcement steel yield strength	Flexure	1.17
	Compression	1.10
	Direct Shear	1.10
	Bond	1.17
Welded wire reinforcement steel yield strength	Flexure	1.10
Hot-rolled steel yield strength	Flexure/Shear	1.19[b]
	Tension/Compression	1.12[b]
Hot-rolled steel ultimate strength	All	1.05[b]
Cold-formed steel yield strength	Flexure/Shear	1.10
	Tension/Compression	1.10

[a]For materials, properties, and failure modes not listed, DIF = 1.00 unless another value is established using a rational procedure that accounts for strain rates and other relevant aspects of element response; assumed strain rates for this table are 0.10 in./in./sec for flexure and shear and 0.02 in./in./sec for tension and compression.
[b]For ASTM A36/A36M steel (ASTM 2008), it shall be permissible to use DIF = 1.29 for yield strength in flexure/shear, DIF = 1.19 for yield strength in tension/compression, and DIF = 1.10 for ultimate strength.
SDOF, single degree of freedom.

TABLE 3-6. DEAD LOAD FACTORS

Element Description	γ_D
Elements in existing buildings for which the actual dead load effect can be established with a high degree of certainty	1.0
All other elements for which the dead load effect reduces the strength available to resist blast effects	1.2
All other elements for which the dead load effect increases the strength available to resist blast effects	0.9

TABLE 3-7. LIVE LOAD FACTORS

Element Description	γ_L
Elements for which risk assessment indicates that the full live load effect is likely to occur simultaneously with the blast effect	1.0
All other elements for which the live load effect reduces the strength available to resist blast effects	0.5
All other elements for which the live load effect increases the strength available to resist blast effects	0

ASTM International
100 Barr Harbor Drive
P.O. Box C700
West Conshohocken, PA 19428-2959
Standard Test Method for Glazing and Glazing Systems Subject to Airblast Loadings, ASTM F1642, 2010.
Standard Test Method for Metal Doors Used in Blast Resistant Applications (Equivalent Static Load Method), ASTM F2247, 2010.
Standard Specification for Carbon Structural Steel, ASTM A36/A36M, 2008.

Chapter 4

BLAST LOADS

4.1 GENERAL

4.1.1 Scope. When deemed appropriate by the procedures specified in Chapter 2, the blast loads shall be determined as specified herein.

4.1.2 Permitted Procedures. The design blast loads for structures shall be determined using the Basic Procedure for External Blast as specified in Section 4.2, the Basic Procedure for Internal Blast as specified in Section 4.3, or Other Procedures as specified in Section 4.4.

4.2 BASIC PROCEDURE FOR EXTERNAL BLAST

4.2.1 Scope. The circumstances under which the blast loads determined in accordance with this section shall be applied to a structure are as follows:

1. The blast occurs on or near the ground surface external to the structure.
2. The blast loading is not significantly affected by surrounding structures or terrain.
3. The structure is a regular shaped structure as defined in Section 1.2.
4. The scaled distance from the blast to the structure is within the scaled ranges of Figs. 4-3 and 4-8.

4.2.2 Directly Loaded Surfaces. The loading shall be idealized as shown in Fig. 4-1 for normal reflection and Fig. 4-2 for oblique reflection.

The equivalent mass of trinitrotoluene (TNT), W_e, shall be determined by multiplying the effectiveness factor from Table 4-1 by the mass of the actual explosive material.

The incident and normally reflected shock wave parameters for the positive phase of the loading shall be determined from Fig. 4-3 in which:

R = minimum distance from detonation to the component being examined
P_{so} = peak side-on or incident overpressure
i_s = incident impulse
t_o = duration
P_r^- = peak normally reflected overpressure
i_r^- = normally reflected impulse
U_s = shock front velocity

The duration of the equivalent triangular incident loading shall be determined from:

$$t_{of} = 2i_s/P_{so} \qquad (4\text{-}1)$$

The duration of the equivalent triangular normally reflected loading shall be determined from:

$$t_{rf} = 2i_r/P_r \qquad (4\text{-}2)$$

The peak dynamic pressure, q_s, shall be determined as a function of the peak side-on overpressure from Fig. 4-4.

The average clearing time to relieve the reflected overpressure shall be determined from:

$$t_c = 4Hw/(w + 2H)C_r \qquad (4\text{-}3)$$

where

H = structure height
w = structure width
C_r = sound velocity from Fig. 4-5

The direct drag load coefficient shall be $C_D = 1$.

The positive phase loading shall be the triangular or bilinear function thus obtained that has the lesser impulse.

For oblique reflections at the incident angle α in Fig. 4-2, the reflected overpressure coefficient, $C_{r\alpha}$, shall be obtained from Fig. 4-6.

The oblique peak reflected overpressure shall be determined from:

$$P_{r\alpha} = C_{r\alpha}P_{so} \qquad (4\text{-}4)$$

The oblique reflected impulse, $i_{r\alpha}$, shall be determined from Fig. 4-7.

The duration of the equivalent triangular obliquely reflected loading shall be determined from:

$$t_{rf} = 2i_{r\alpha}/P_{r\alpha} \qquad (4\text{-}5)$$

If the negative phase of the loading is to be considered, the parameters shall be determined from Fig. 4-8 in which:

$P_{r'}$ = peak normally reflected pressure
$i_{r'}$ = normally reflected impulse.

The duration of the equivalent triangular negative phase loading shall be determined from:

$$t_{rf}^- = 2\frac{i_r^-}{P_r^-} \qquad (4\text{-}6)$$

The rise time of the negative phase loading shall be $t_{rf}^-/4$ as shown in Figs. 4-1 and 4-2.

4.2.3 Indirectly Loaded Surfaces. The loading shall be the linear function, $P(t) = P_s + C_D q$, shown in Fig. 4-1, in which P_s is the overpressure and q the dynamic pressure. The parameters of this function shall be obtained as described in Section 4.2.2. However, for these indirectly loaded roof, side wall, and rear wall surfaces, the drag coefficient shall be the negative values obtained from Table 4-2.

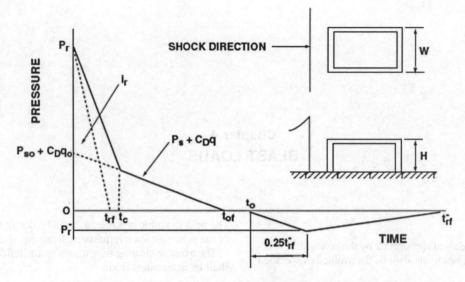

FIGURE 4-1. FRONT WALL LOADING, NORMAL REFLECTION.

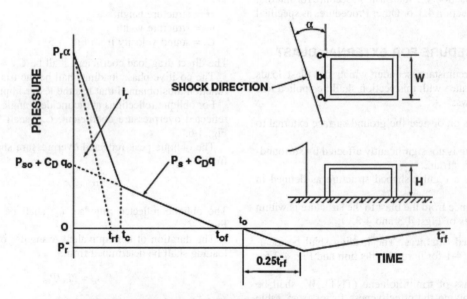

FIGURE 4-2. FRONT WALL LOADING, OBLIQUE REFLECTION.

TABLE 4-1. EQUIVALENT TNT MASSES FOR AIRBLAST IN FREE AIR

Explosive	Density (lb/ft³)	Equivalent Mass for Pressure	Equivalent Mass for Impulse	Overpressure Range (psi)
Amatol (50/50)	99	0.97	0.87	NA[a]
Ammonia dynamite (50% strength)	NA[a]	0.90	0.90[b]	NA[a]
Ammonia dynamite (20% strength)	NA[a]	0.70	0.70[b]	NA[a]
ANFO (94/6 ammonium nitrate/fuel oil)	NA[a]	0.87	0.87[b]	4 to 1,000
AFX-644	109	0.73[b]	0.73[b]	NA[a]
AFX-920	99	1.01[b]	1.01[b]	NA[a]
AFX-931	101	1.04[b]	1.04[b]	NA[a]
Composition A-3	103	1.09	1.07	4 to 51
Composition B	103	1.11 1.20	0.98 1.30	4 to 51 100 to 1,000
Composition C-3	100	1.05	1.09	NA[a]

TABLE 4-1. EQUIVALENT TNT MASSES FOR AIRBLAST IN FREE AIR (continued)

Explosive	Density (lb/ft³)	Equivalent Mass for Pressure	Equivalent Mass for Impulse	Overpressure Range (psi)
Composition C-4	99	1.20 1.37	1.19 1.19	10 to 200
Cyclotol (75/25 RDX/TNT) (70/30) (60/40)	107 108 109	1.11 1.14 1.04	1.26 1.09 1.16	NA[a] 4 to 51 NA[a]
DATB	112	0.87	0.96	NA[a]
Explosive D	107	0.85[b]	0.81	1 to 44
Gelatin dynamite (50% strength)	NA[a]	0.80	0.80[b]	NA[a]
Gelatin dynamite (20% strength)	NA[a]	0.70	0.70[b]	NA[a] .
H-6	110	1.38	1.15	4 to 102
HBX-1	110	1.17	1.16	4 to 20
HBX-3	115	1.14	0.97	4 to 25
HMX	NA[a]	1.25	1.25[b]	NA[a]
LX-14	NA[a]	1.80	1.80[b]	NA[a]
MINOL II	114	1.20	1.11	3 to 20
Nitrocellulose	103 to 106	0.50	0.50[b]	NA[a]
Nitroglycerin dynamite (50% strength)	NA[a]	0.90	0.90[b]	NA[a]
Nitroguanidine (NO)	107	1.00	1.00[b]	NA[a]
Nitromethane	NA[a]	1.00	1.00[b]	NA[a]
Octol (75/25 HMX/TNT) (70/30)	113 88	1.02 1.09	1.06 1.09[b]	NA[a] 1 to 44
PBX-9010	112	1.29	1.29[b]	4 to 31
PBX-9404	113	1.13 1.70	1.13[b] 1.70	4 to 100 100 to 1,000
PBX-9502	118	1.00	1.00	NA[a]
PBXC-129	107	1.10	1.10[b]	NA[a]
PBXN-4	107	0.83	0.85	NA[a]
PBXN-107	102	1.05[b]	1.05[b]	NA[a]
PBXN-109	104	1.05[b]	1.05[b]	NA[a]
PBXW-9	NA[a]	1.30	1.30[b]	NA[a]
PBXW-125	112	1.02[b]	1.02[b]	NA[a]
Pentolite (cast)	102 105 NA[a]	1.42 1.38 1.50	1.00 1.14 1.00	4 to 100 4 to 600 100 to 1,000
PENT	110	1.27	1.27[b]	4 to 100
Picrotol (52/48 Ex D/TNT)	102	0.90	0.93	4 to 595
RDX	NA[a]	1.10	1.10[b]	NA[a]
RDX/Wax (98/2)	120	1.16	1.16[b]	NA[a]
RDX/AL/Wax (74/21/5)	NA[a]	1.30	1.30[b]	NA[a]
TATB	NA[a]	1.00	1.00[b]	NA[a]
Tetryl	108	1.07	1.07[b]	3 to 20
Tetrytol (75/25 Tetryl/TNT)	99	1.06	1.06[b]	NA[a]
TNETB	106	1.13	0.96	4 to 100
TNETB/Al (90/10) (78/22) (65/35)	109 74 77	1.23 1.32 1.38	1.11 1.32[b] 1.38[b]	4 to 100 NA[a] NA[a]
TNT	102	1.00	1.00	Standard
Tropex	115	1.23	1.28	1 to 44
Tritonal (80/20 TNT/AL)	107	1.07	0.96	4 to 100

[a]NA, data not available.
[b]Value is estimated.

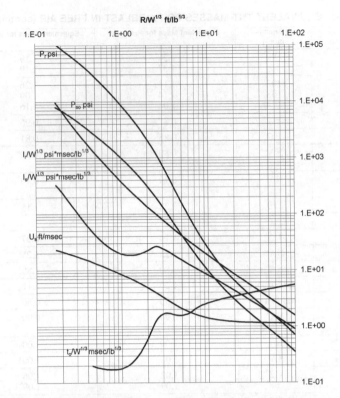

FIGURE 4-3. POSITIVE PHASE SHOCK PARAMETERS FOR EXPLOSIONS OF HEMISPHERICAL TNT CHARGES ON THE SURFACE AT SEA LEVEL.

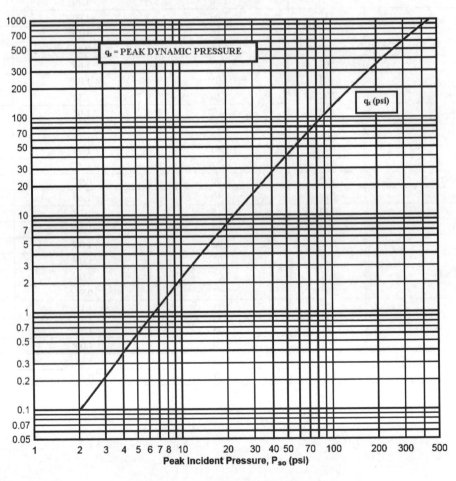

FIGURE 4-4. PEAK DYNAMIC PRESSURE.

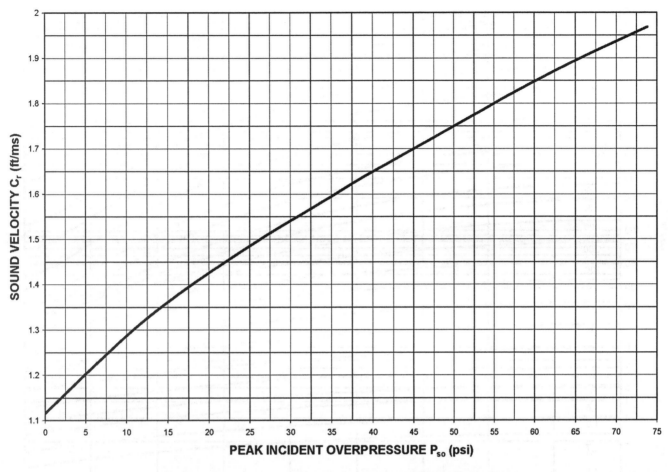

FIGURE 4-5. SOUND VELOCITY IN REFLECTED OVERPRESSURE REGION.

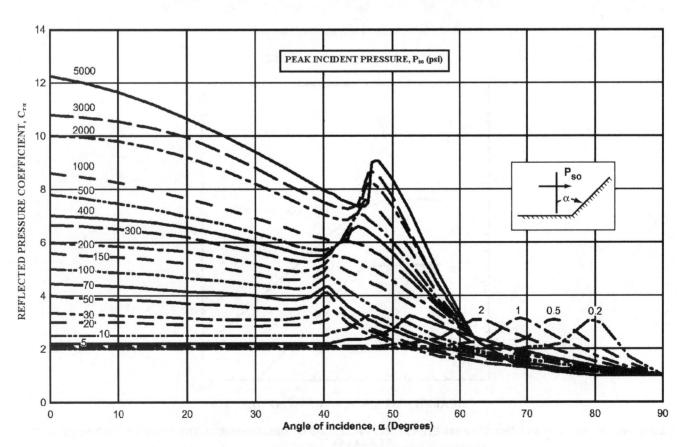

FIGURE 4-6. REFLECTED OVERPRESSURE COEFFICIENT.

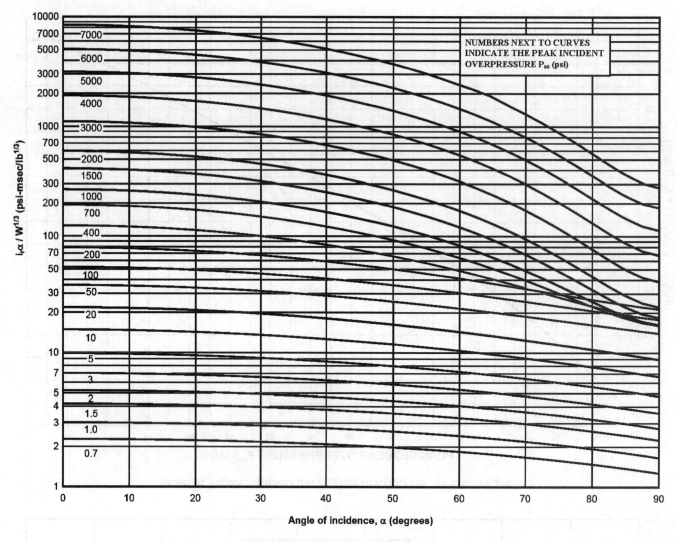

NUMBERS NEXT TO CURVES
INDICATE THE PEAK INCIDENT
OVERPRESSURE P_{so} (psi)

Angle of incidence, α (degrees)

FIGURE 4-7. REFLECTED IMPULSE.

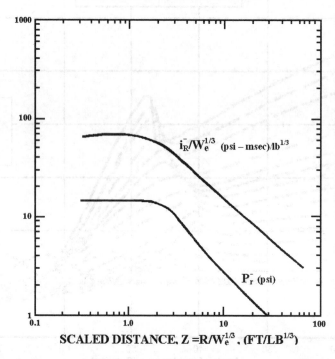

$i_R^-/W_e^{1/3}$ (psi – msec)/lb$^{1/3}$

P_r^- (psi)

SCALED DISTANCE, Z = R/W$_e^{1/3}$, (FT/LB$^{1/3}$)

FIGURE 4-8. NEGATIVE PHASE SHOCK PARAMETERS FOR EXPLOSIONS OF HEMISPHERICAL TNT CHARGES ON THE SURFACE AT SEA LEVEL.

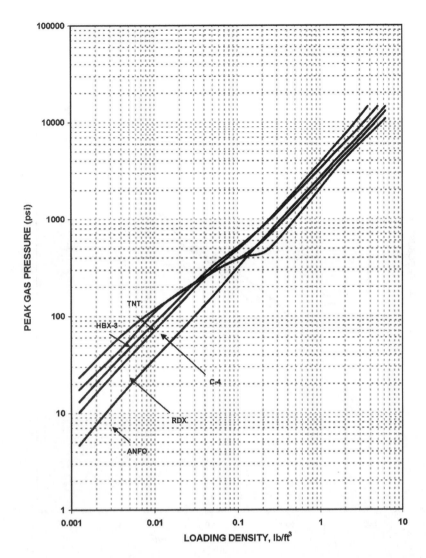

FIGURE 4-9. PEAK GAS OVERPRESSURE FOR RDX, TNT, C-4, HBX-3, AND ANFO.

<table>
<tr><td colspan="2">TABLE 4-2. SIDE-ON ELEMENT DYNAMIC DRAG COEFFICIENTS</td></tr>
<tr><td>Peak Dynamic Pressure
q_o</td><td>Drag Coefficient
C_D</td></tr>
<tr><td>0 to 25 psi</td><td>−0.40</td></tr>
<tr><td>25 to 51 psi</td><td>−0.30</td></tr>
<tr><td>51 to 145 psi</td><td>−0.20</td></tr>
</table>

4.3 BASIC PROCEDURE FOR INTERNAL BLAST

4.3.1 Scope. The circumstances under which the blast loads determined in accordance with this section shall be applied to a structure are as follows:

1. The blast occurs internal to the structure.
2. The structure is a regular shaped structure as defined in Section 1.2 and has a maximum aspect ratio less than 1.3 in plan area.

3. The minimum covered vent area, A_v, is greater than or equal to $0.20 \, V_f^{2/3}$ where V_f is the internal free volume of the structure.
4. The unit weight of the vent(s) is less than or equal to 25 psf.

4.3.2 Procedure. The peak gas overpressure, P_g, shall be determined from Fig. 4-9 or 4-10 for the particular explosive compound. The interior surface of interest shall be designed for this gas overpressure applied statically. It shall be permitted to design by another procedure acceptable to the Authority Having Jurisdiction.

4.4 OTHER PROCEDURES

Structures not meeting the requirements of Section 4.2.1 or 4.3.1 shall be designed using recognized literature documenting such blast-load effects. The recognized literature shall be permitted in lieu of the Basic Procedures for any structure.

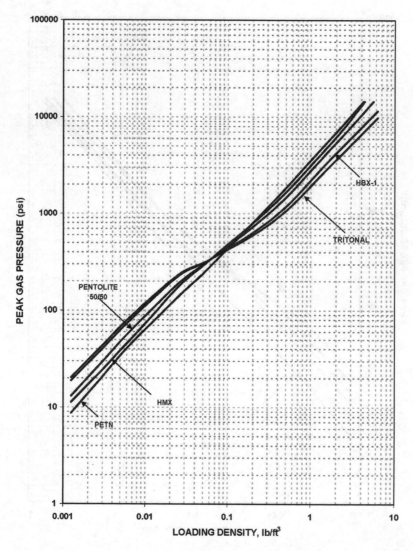

FIGURE 4-10. PEAK GAS OVERPRESSURE FOR HMX, PETN, PENTOLITE, HBX-1, AND TRITONAL.

Chapter 5
FRAGMENTATION

5.1 GENERAL

5.1.1 Scope. When deemed appropriate by the project-specific risk assessment, the building or specific areas within the building shall be designed to mitigate the effects of primary or secondary fragments.

5.2 DESIGN REQUIREMENTS

Design for fragmentation shall meet acceptable performance conditions determined by the project-specific risk assessment. Acceptable performance conditions shall be specified as: (1) no penetration, or (2) penetration without perforation. For concrete and masonry elements, acceptable performance conditions shall address direct spalling on the protected side during an airblast.

5.3 ANALYTICAL PROCEDURES

Fragmentation analysis for structures shall conform to the requirements of this section.

5.3.1 Acceptable Analytical Methods. Design for fragment impact effects shall be in accordance with methods deemed acceptable by the Authority Having Jurisdiction, as defined in Section 1.2.

5.3.2 Limits on Analytical Procedures. Analytical procedures for predicting the behavior of fragments are considered approximate due to the uncertain nature of the threat, fragments, and site surrounding a commercial building.

5.3.3 Complex Modeling Methods. When a complex modeling method such as finite element analysis or computational fluid dynamics is used for determining fragment velocity, flight, or impact damage, peer review shall be required in accordance with Section 10.2.

Chapter 6
STRUCTURAL SYSTEMS

6.1 GENERAL PROVISIONS

6.1.1 Purpose. This chapter presents methods for the design and construction of buildings and similar structures subject to the effects of blast loads. The structure's response to blast loads and the structure's resistance to progressive collapse may involve dissipation of energy through inelastic deformation. The provisions for design and construction presented in this chapter shall be satisfied even when load combinations that do not contain the blast effect indicate larger demands than the combinations including blast.

6.1.2 Scope and Application. The provisions of this chapter apply to new structures and to additions to existing structures and structural renovations when required by the Authority Having Jurisdiction. The principles presented in this chapter address near-range (close-in) explosions with a scaled distance less than 3.0 ft/lb$^{1/3}$ (1.2 m/kg$^{1/3}$) and far-range explosions relative to the structure, as defined in Chapter 1. Explosive effects of charges in contact or near contact with the structural elements, which require testing or detailed computational fluid dynamics and explicit finite element analyses, are discussed in Section 6.4.1. The analysis and design of structures in response to blast loading are distinct from seismic design and must be considered independently.

6.2 STRUCTURAL MODELING AND ANALYSIS

6.2.1 Analytical Methods. The analysis of structures to resist explosive loads shall be performed using one of the four methods presented in Sections 6.2.1.1, 6.2.1.2, 6.2.1.3, or 6.2.1.4. See Section 3.3.2 for definitions of element behavior for different levels of protection (LOPs).

6.2.1.1 Pressure-Impulse Charts. Pressure-impulse (P-I) charts shall either be generated using the general principles described in this Standard or by approved testing in accordance with any of the following: (1) government test data acquisition procedures; and/or (2) test data collection in accordance with Chapter 10 of this Standard. The use of P-I charts shall be limited to flexural modes in response to external blast loads and to structural elements that are within the P-I database. Diagonal tension and direct shear capacity must be verified independently.

6.2.1.2 Single Element Response Analysis. It shall be permissible to analyze primary and secondary structural elements independently of each other by means of either single-degree-of-freedom (SDOF) or multi-degree-of-freedom (MDOF) inelastic dynamic methods where the expected failure mechanism can be adequately represented by the simplified dynamic system.

It shall be permissible to include the effects of structural damping in SDOF or MDOF analyses, particularly when the response is nearly elastic.

6.2.1.3 Structural System Multi-Degree-of-Freedom Finite Element Response Analysis. MDOF finite element modeling of structural systems that accounts for the interaction between blast loading and the response of interconnected elements, the phasing of their responses, the flexibility of the actual boundary conditions, and the inelastic behavior of the elements shall be used when (1) such effects cannot be neglected or conservatively accounted for in the methods described in Sections 6.2.1.1 and 6.2.1.2, and (2) where the expected failure mechanism can be adequately represented by the MDOF finite element system.

6.2.1.4 Explicit Linear or Nonlinear Finite Element Analysis. Advanced analytical methods shall be used for primary structural systems when the design professional determines that inadequacies in simplified approximations reduce the reliability of the calculated results or where criticality of primary systems (such as transfer systems) requires greater reliability. Explicit dynamic linear or dynamic nonlinear finite element models shall be used to analyze the structural response of structures and structural elements to blast loading when the methods described in Sections 6.2.1.1, 6.2.1.2, and 6.2.1.3 cannot be used to model adequately the response of the structural system. Explicit nonlinear finite element analysis shall be required when close-in effects are of primary concern and are required by Section 6.4.1. Explicit nonlinear finite element analysis shall also be required when temporal or spatial distribution of loading and spatial variation in mass or structural properties cannot be represented by simplified methods. The results of the explicit nonlinear finite element analysis shall demonstrate the performance of the structure and its ability to continue to carry gravity loads concurrent with the dynamic response. The response limits applied to nonlinear explicit dynamic finite element analyses shall satisfy the performance requirements of Section 3.3, where applicable. The SDOF response limits of Section 3.4 shall not be applied.

6.2.2 Materials. Characterizations of structural elements for modeling purposes shall represent both the elastic and inelastic behavior of the materials. For the modeling of flexural elements, an elastic-perfectly-plastic resistance function shall be the simplest analytical representation. The elastic-perfectly-plastic plateau shall not exceed the average of the materials' dynamic yield strength and dynamic ultimate strength. The different tensile and compressive plateaus shall be considered when the structural element is exposed to combined in-plane membrane limit loads concurrent with flexural limit loads. Chapter 9 provides guidance on the use of limit state properties for materials commonly used in construction.

6.2.2.1 Material Strength Increase. It shall be permissible to use average strength factors (ASF) and dynamic increase factors (DIF) in accordance with Section 3.5 for SDOF and MDOF analyses. The expected dynamic design strength shall be determined to be one-half the sum of the expected dynamic yield and expected dynamic ultimate strengths, according to Section 3.5.1, of the materials undergoing large deformations. Average strength and dynamic increase effects for explicit dynamic linear or non-linear finite element analyses shall be represented by a validated constitutive model for each of the materials.

6.2.3 Modeling of Elements. In addition to flexural modes of response, structural analysis shall account for direct shear, axial interaction with flexure, diagonal tension shear, and brittle failure modes, such as the fracture of welds and bolts and brittle connection failure.

6.2.3.1 Flexure. Bending moments in elements subject to blast loading shall be determined by an analysis method that considers the material strain rate produced by the blast loading, the dynamic response of the system, the elastic and inelastic materials behavior of the elements, and permissible deformation and ductility limits.

Ductile flexural elements shall perform in the inelastic response range undergoing large displacements without brittle failure due to shear, fracture, or failure due to instability. Structures that require LOP IV (a high level of protection, and are therefore designed to remain nearly elastic in response to design level threats) shall also be detailed to develop larger levels of ductility in response to a greater-than-design-level event. The effects of rebound shall be considered in the design and detailing of all flexural elements.

6.2.3.1.1 Flexure in Structural Steel Elements. Steel elements shall satisfy the requirements of Section 9.3 for the different LOPs.

Splices in elements are not permitted in regions where plastic hinges are expected to occur. Full moment connections, where required, shall be stronger than the maximum plastic moment that can be developed when considering the sources of material average strength and dynamic strength increase effects identified in Section 6.2.2.1. Brittle fracture failure modes in the connections shall be avoided and the governing failure mode of the connections shall be yielding of steel elements, such as plates, instead of fracture of bolts or welds. Adequate lateral bracing shall be provided at all plastic hinge locations. Inflection points shall not be permitted to serve as points of lateral restraint for plastic hinges.

6.2.3.1.2 Flexure in Concrete Elements. Unless a more rigorous form of analysis is used, such as an explicit finite element analysis approach, or full scale testing, the determination of the dynamic response of a reinforced concrete element subject to blast loading in the elastic and elastoplastic ranges shall be based upon a dynamic analysis, where the moment of inertia of the element is recommended to be the average of the gross moment of inertia and the cracked moment of inertia. Use of an alternate inertia shall be based on a rational defendable design approach.

6.2.3.1.3 Flexure in Masonry Elements. Wall supports shall develop full flexural capacity of the wall. Axially loaded walls shall be designed considering slenderness effects.

6.2.3.2 Shear, Axial, and Reaction Forces in Bending Elements. All flexural elements and their connections shall be designed and detailed such that no brittle failure mode becomes the limiting failure mode. Unless the element is designed to remain elastic under blast loading, ductile failure modes shall be the governing failure mode for flexural elements and their connections and splices. If the elements are designed to resist the blast loads elastically, the design of nonductile modes shall include a 1.5 factor of safety on the calculated forces.

Shear, axial, and reaction forces used for the design of elements subject to flexure loading due to blast effects shall be determined by a capacity analysis considering material strength increase factors identified in Section 6.2.2.1. The design shear forces shall not be less than the shear forces associated with the nominal flexural strength of the element.

Concrete elements subjected to flexural and axial loads in excess of one-tenth the product of the gross cross-sectional area and the specified compressive strength shall be designed and detailed as elements in combined flexure and compression.

6.2.3.2.1 Shear in Structural Steel Elements. The design shear forces shall not be less than the shear forces associated with the ultimate flexural strength of the element, unless the shear force is demonstrated by explicit finite element analysis or if elements are designed to resist the blast loads elastically. Flexural analysis of steel elements supporting concrete slabs shall take into account composite action, where shear connectors are provided, to determine the end reactions for connection design.

6.2.3.2.2 Shear in Reinforced Concrete Elements. The design shear forces shall not be less than the shear forces associated with the nominal flexural strength of the element, unless the shear force is demonstrated by explicit finite element analysis or if elements are designed to resist the blast loads elastically. Ultimate flexural resistance for determining element shear shall not be less than the ultimate flexural resistance of the element when analyzed as a Type I cross section according to UFC 3-340-02 (DoD 2008).

The direct shear force shall be determined at the face of the support.

For external blast loads, it shall be permissible to compute the diagonal tension shear force at a distance from the support equal to the effective depth of the element, provided that the element is loaded uniformly. Shear reinforcing shall be provided for a distance at least the effective depth further than that needed by analysis (while satisfying minimum shear reinforcement requirements) and the element is supported by bearing and there are no concentrated loads on the element within a distance from the support equal to the effective depth. For internal blast loads, the diagonal tension shall be computed at the face of the support. When axial loads due to blast loads result in net tension in reinforced concrete elements, then the shear capacity of the concrete, V_c, shall be neglected.

Flexural elements with a span-to-depth ratio less than 4 shall be analyzed as deep beams accounting for the nonlinearity of strain distribution over the depth of the cross section and shall be detailed in accordance with Chapter 9.

6.2.3.2.3 Shear in Masonry Elements. The design shear forces shall not be less than the shear forces associated with the nominal flexural strength of the element. Masonry elements that are designed to resist blast loads and are reinforced only within the horizontal bed joints shall utilize shear reinforcement designed to carry the total shear stress. Alternatively, deeper wall sections shall be used. Shear reinforcement shall be detailed in accordance with Chapter 9.

6.2.3.3 Axial Load Effects. Tension and compression shall be considered in combination with other structural actions in accordance with the following subsections.

6.2.3.3.1 Compression Elements. The design of ductile steel, reinforced concrete, and reinforced masonry compression elements with or without flexure shall include the amplified bending moments due to second-order (i.e., P-delta) effects.

Shear forces in ductile flexural elements subject to static axial compression loads and blast axial compression loads, in accordance with Section 3.5.3, shall be determined based on the requirements for flexural elements in Section 6.2.3.2.1, 6.2.3.2.2, or 6.2.3.2.3. If axial loading increases the flexural resistance of the element, then axial loads on the element shall be included when determining the flexural resistance for shear. For ductile flexural elements with compression, the static axial loading shall be based on actual dead and live loads. Dynamic blast compression load shall be as determined by blast analysis.

Analysis shall consider the flexural effects in compression elements due to joint fixity.

6.2.3.3.2 Tensile Elements. The required strength of connections and splices in tension elements subject to blast loads shall equal the actual dynamic tension strength of the element. It is permissible to include the material's over-strength, in accordance with Section 6.2.2.1. Appropriate average strength and dynamic increase effects can also be used for components of the tension connections as given in Chapter 3.

6.2.3.4 Instability. Instability of the overall structure, or a significant portion of it, or premature buckling of individual elements shall be considered. The global effect of the formation of plastic hinges shall be evaluated. At a minimum, a global stability check shall be performed on the damaged state, resulting from the explosive event, to verify the effect of the formation of plastic hinges.

6.2.3.4.1 Element and System Stability. Analysis shall consider overall stability of the structural system resulting from the following:

1. Buckling of elements and systems
2. Frame side-sway and P-delta effects
3. Loss of diaphragms
4. Loss of bearing.

6.2.3.4.2 Local and Lateral-Torsional Buckling. Analysis shall consider local buckling and bracing for individual elements that are required to undergo large deformations.

6.2.3.4.3 Instability Initiated by Damage. Analysis shall consider instability that results from damage to structural elements and systems due to blast pressures and fragments. It shall include the following:

1. Loss of or damage to primary bracing elements
2. Loss of diaphragm capacity
3. Fragments impacting and damaging the structure
4. Plastic hinging
5. Progressive collapse.

6.2.3.4.4 Progressive Collapse. Ductile modes of structural failure (yielding, deflection, etc.) in response to extreme loading events are acceptable if the desired LOP is satisfied.

All buildings three stories or more and designed in accordance with this Standard shall be designed to prevent vertical progressive collapse resulting from damage caused by an explosive event. The structure shall either be designed to prevent the loss of critical load-bearing elements, or the structure shall be designed to bridge over the damaged area resulting from the blast effects as defined by a risk assessment. The prevention of progressive collapse shall incorporate either indirect design approaches, such as tie-force methods, or direct design approaches, such as alternate path or local hardening methods, as deemed appropriate by the responsible design professional. The evaluation for the potential of progressive collapse shall take into account the longer time regime for collapse to propagate as compared to the structure's response to blast loading.

Buildings of all heights shall be designed to prevent a horizontal progressive collapse due to the loss of lateral resistance.

The design to prevent progressive collapse shall be applied to all buildings, except when exempted by the Authority Having Jurisdiction or where the threat assessment permits the alternative hardening to resist a specified explosive threat.

6.2.4 Connections and Joints. Connections and joints shall be designed and detailed to resist axial and shear forces, bending moment, and torsion that must be transferred to supporting elements. The effects of rebound shall be considered in the design and detailing of all connections. Connections and joints that are designed for elastic response shall include a 1.5 factor of safety for nonductile modes, but not greater than the maximum reaction force that can be transferred by the element.

6.2.5 Application of Loads. Blast loads on individual structural elements shall be as specified in Chapter 4. It shall be permissible to apply blast load to the dynamic representation of the system using a range of behavioral assumptions. Equivalent uniform loads applied to SDOF models shall be determined by a rational method. It shall be permissible to use energy methods to calculate an equivalent uniform load in lieu of the actual applied forces for each interval of a time-varying load function. In this approach, the total work performed by an actual load combination shall be equal to the work performed by an equivalent uniform load. However, for all analytical approaches, the total impulse of the blast effects to be applied over the loaded surface shall be preserved.

6.2.5.1 Empirical Data, Ray Tracing, and Computational Fluid Dynamics. It shall be permissible to use empirical data for most blast loading calculations. Reflections that may result from multiple surfaces, such as floor and ceiling slabs relative to walls, shall be considered using ray tracing methods or appropriate reflection factors. Advanced analytical methods, such as computational fluid dynamics (CFD) analyses, shall be used for very short standoff distances, complex geometries consisting of multiple reflecting surfaces, and confined spaces.

6.2.5.2 Spatial Distribution. The blast load applied to each structural element shall be representative of the distribution of pressures over its surface area. Short-standoff detonations are associated with large gradients of peak pressures and the representative blast loading shall be applied in a suitable fashion to capture the flexural and direct shear modes of response. The spatial distribution of blast loads shall be conservatively enveloped by considering the practical range of standoff distances associated with the design basis threat.

6.2.5.3 Temporal Distribution. The variation of blast loads over time shall be as specified in Chapter 4. It shall be permissible to approximate this variation as straight-line segments (linear decay), provided that the impulse over the duration of the pulse is preserved. Multiple pulses, resulting from reflections off of adjacent surfaces, shall be phased to preserve the arrival times and superposed with the primary shock wave. Similarly, the gas pressures associated with detonations within confined spaces shall be superposed with the shock wave loading. The temporal distribution of blast loads shall be conservatively enveloped by considering the practical range of standoff distances associated with the design basis threat.

6.2.5.4 Negative Phase. The negative phase that follows a shock front is specified in Section 4.2.2. The negative phase shall only be included in the superposition of blast pulses if multiple reflections are accurately considered and the spatial distribution of loads are accurately represented using CFD analytical methods. Scenarios in which the influence of the negative phase does not conservatively envelope the blast loading shall not be considered.

The negative phase shall be considered when the effects resulting from this component of loading create a critical condition or where negative pressures are in phase with rebound motions.

6.2.5.5 Element-to-Element Load Transfer. Concentrated blast loads, provided by the reactions of secondary structural elements that frame into the structural element under consideration, shall be addressed in a consistent manner. Either multi-degree-of-freedom models shall be used to apply the representative spatial distribution or "equivalent" loads, derived by conservation of energy methods, shall be applied to simplified SDOF models.

6.2.5.6 Tributary Loads. All loads tributary to the structural element being analyzed shall be considered. It shall be permissible to apply these loads to SDOF models as equivalent uniform loads with the mass associated with the collecting surfaces assumed to respond with the dynamic system. It shall also be permissible to apply these loads to SDOF models as reaction force time-histories from separate SDOF models of the secondary elements that frame into the structural element. However, the appropriate mass of the secondary element shall be that which is assumed to respond with the dynamic system.

6.2.5.7 Load Combinations. Blast loads shall be applied in combination with other loads that may be acting on the element at the time of the explosion in accordance with Section 3.5.3. In moderate and high seismic areas where seismic detailing is utilized, special consideration shall be given to structural detailing to ensure that designated structural systems are detailed to resist both events.

6.2.5.8 Clearing Effects. Blast pressures that are applied to the surfaces of buildings or building components shall account for clearing effects in the presence of an edge or opening. See Section 4.2.2.

6.2.6 Mass. Inertial properties of dynamic systems shall be determined in accordance with established methods of structural dynamics. The mass used in the analysis shall be limited to the mass of the structure; however, additional nonstructural mass may be used if it will stay intact during the blast. SDOF methods shall incorporate the appropriate mass factors associated with the respective structural elements and response characteristics.

6.3 STRUCTURAL DESIGN

6.3.1 Structural Systems. Structural framing and load-bearing elements shall have adequate strength and ductility to resist the specified loads. Primary structural systems and secondary structural systems shall be designed and detailed to achieve their respective performance expectations for the specified LOP in accordance with Section 3.3.

6.3.1.1 Steel Moment Frame Systems. Except as provided in Section 9.3.2.5, compact sections shall be used for individual elements, in accordance with Section 9.3, in order for plastic moments to develop as determined in the nonlinear dynamic analyses. Connections in moment frames shall be designed to develop the maximum reaction forces, both shear and axial forces, and moments for the individual elements. Consideration shall be given to designing the connections to resist the fully inelastic reaction forces and moments for blast loading in which individual elements are found to remain elastic. Overall frame stability, including the combined effects of axial load and lateral deformations, shall be determined, in accordance with Section 6.2.3.4, for conditions where plastic hinges are allowed to develop.

6.3.1.2 Steel Braced Frame Systems. Except as provided in Section 9.3.2.5, compact sections shall be used for individual elements, in accordance with Section 9.3, in order for plastic moments to develop as determined in the nonlinear dynamic analyses. Element connections shall be designed to develop the maximum reaction forces for the individual elements. Consideration shall be given to designing the connections to resist the fully inelastic reaction forces for blast loading in which individual elements are found to remain elastic. Overall frame stability shall be determined, in accordance with Section 6.2.3.4, for conditions where plastic hinges are allowed to develop.

6.3.1.3 Concrete Moment Frame Systems. Concrete moment frame elements shall be adequately confined and the shear capacity shall be sufficiently developed in order for individual elements to achieve their plastic capacity and develop ductile inelastic deformations. The combined effects of flexure and axial loads for the columns shall be evaluated in response to the specified blast loads. Overall frame stability shall be determined for conditions where plastic hinges are allowed to develop.

6.3.1.4 Concrete Frame with Concrete Shear Wall Systems. Concrete frame elements shall be adequately confined and the shear capacity shall be sufficiently developed in order for individual elements to achieve their plastic capacity and develop ductile inelastic deformations. The combined effects of flexure and shear in the wall shall be evaluated in response to the specified blast loads. Overall structural stability shall be determined for conditions where inelastic deformations are allowed to develop. Walls shall be designed and detailed for in-plane and out-of-plane response to direct airblast loadings.

6.3.1.5 Precast Tilt-Up with Concrete Shear Wall Systems. Panelized systems shall be adequately tied together to enable individual elements to develop allowable deformations without loss of bearing or load transfer capacity, in response to both direct blast loading and rebound. Connections shall be designed, detailed, and constructed such that the governing failure mode of the connection is a ductile mode related to yielding of steel elements of the connection such as angles, plates, rebar, and anchors. Fracture of bolts and welds shall not be the governing failure mode of connections.

6.3.1.6 Reinforced Masonry Bearing/Shear Walls. Masonry wall structures shall be sufficiently reinforced in order to resist the in-plane and out-of-plane response to dynamic blast loads. It shall be permissible to consider arching action for the evaluation of wall response and use empirical P-I charts based on masonry wall tests. Although the impulse asymptote "layover" associated with the beneficial effects of negative phase loading may be considered for evaluation of existing masonry walls, the reliance on negative phase shall not be considered for design. Diagonal tension and direct shear capacity must be verified independently. Walls shall be designed to develop the shear forces associated with plastic hinging. Connections shall be designed, detailed, and constructed such that the governing failure mode of the connection is a ductile mode related to yielding of steel elements of the connection such as angles, plates, rebar and anchors. Fracture of bolts and welds shall not be the governing failure mode of connections.

6.4 RESPONSE CHARACTERISTICS

6.4.1 Close-In Effects. It shall be permissible to analyze structural components subjected to unconfined close-in blast effects for the impulse rather than the peak pressure. This applies to conditions in which the pulse duration is short relative to the natural period of the structural element. Structural elements required to withstand interior close-in explosion effects shall be designed for shock impulses and the gas pressures. Design performance in response to close-in effects shall be determined by the owner.

When designing blast protective structures for close-in effects, breach of the structure, direct shear, flexure, diagonal tension, and spall shall be considered as failure modes. For LOP IV (high level of protection), except as provided otherwise herein, test data or CFD analysis shall be performed to determine blast loading associated with short-standoff scaled ranges, and explicit nonlinear finite element analyses shall be performed to determine the associated effects of breach, direct shear, diagonal tension, and spall. Simplified analyses and empirical relations shall not be used to determine close-in effects other than to estimate gross characteristics for lower LOPs or when the empirical relations have been validated by test or physics-based analyses that model similar threat scenarios and construction details.

6.4.2 Far-Range Effects. Structures subjected to blast effects in the far range shall be designed in response to a uniformly loaded plane wave in accordance with Section 6.2.

6.5 CONSENSUS STANDARDS AND OTHER REFERENCED DOCUMENTS

References cited in the Commentary to Chapter 6, but not given in Section C6.5, include the following:

ACI
American Concrete Institute
38800 Country Club Drive
Farmington Hills, MI 48331
Building Code Requirements for Structural Concrete (ACI 318-08) and Commentary (ACI 318R-08), 2008.
Steel Construction Manual, 13th ed.
Specifications for Structural Steel Buildings, ANSI/AISC 360-10.
Seismic Provisions for Structural Steel Buildings, ANSI/AISC 341-05.

ASCE
American Society of Civil Engineers
1801 Alexander Bell Drive
Reston, VA 20191-4400
Minimum Design Loads for Buildings and Other Structures, ASCE/SEI 7-05.

U.S. Department of Defense
Structures to Resist the Effects of Accidental Explosions, UFC 3-340-02, 2008.

Chapter 7
PROTECTION OF SPACES

7.1 GENERAL PROVISIONS

Structural elements of buildings specifically intended to protect the occupants of the rest of the structure from isolated threats shall be designed and constructed in accordance with the provisions specified herein. Section 7.2 addresses the design of buildings with controlled access and thus applies only to the design of commercial and private buildings with an appropriate level of security. For buildings that allow uncontrolled and unmonitored entry, threats may be located anywhere and the provisions of Section 7.2 are not mandatory. Section 7.3 provides design requirements for safe havens.

7.2 WALLS AND SLABS ISOLATING INTERNAL THREATS

7.2.1 Scope. This section applies to all buildings with controlled access and addresses the design of structural elements that protect occupants of the rest of the structure from isolated internal threats.

7.2.2 Threat Locations for Buildings with Controlled Access. For buildings with controlled access, internal explosive threats shall be considered at the locations where they may be introduced. Such locations include, but are not limited to, loading docks, mailrooms, and designated screening rooms. These and other locations shall be determined as part of the risk assessment process described in Chapter 2. For new construction, access control and screening shall prevent explosive threats from entering beyond the building perimeter.

7.2.3 Design Provisions for Walls and Slabs Isolating Internal Threats. Identified threat locations, walls, slabs, and roof construction (if applicable) that form the isolation shell shall be designed and detailed in accordance with Chapter 6. The magnitude of the explosive and its critical locations to be considered in the design of these elements are case-specific and shall be determined as part of the risk assessment process described in Chapter 2. With appropriate consideration of its relative location, the design magnitude of the explosive, and the construction of the surrounding structure, the isolation room shall be designed to withstand the venting of blast pressures from the explosive source as calculated in accordance with Chapter 4. The performance in response to internal detonations shall be determined in accordance with Section 6.4.1 where close-in effects are applicable.

7.2.4 Stairwell Enclosures. The design pressures on the stairwell enclosure walls due to an external blast shall be determined in accordance with the risk assessment as described in Chapter 2. These enclosures shall be designed and detailed to resist the pressure and impulse calculated in accordance with Chapter 4. Unless CFD analytical methods are used to calculate blast pressures on stairwell enclosures inside the building perimeter, the blast loads shall conservatively neglect shielding from the attached structure. For new construction, stairwell enclosures shall not be located in close proximity to isolated threat locations unless it can be demonstrated that full building evacuation can be accomplished within an acceptable time despite the loss of a single stairwell.

7.2.5 Hardened Plenums. This section applies only to mechanical equipment requiring specific blast protection as identified during the risk assessment process outlined in Chapter 2.

7.2.5.1 Design Provisions for Hardened Plenums. For mechanical equipment requiring blast protection, structural elements forming the mechanical plenum that surrounds the equipment shall be hardened to resist the reflected pressures and impulses without fragmentation or dismemberment that would otherwise cause debris to be propelled into the protected space. The use of blast valves shall be considered where plenums cannot be constructed or the blast infill pressures cannot be reduced to acceptable levels.

7.3 SAFE HAVENS

Safe havens are not mandatory unless required by the Authority Having Jurisdiction. The major properties related to the design of safe havens include type, location, and design criteria. After determining these properties, the design of a safe haven shall proceed using the information provided in other sections of this Standard.

7.3.1 Design Considerations. Safe havens shall be enclosed with blast-resisting partitions and within robust structural systems that will resist collapse. The safe haven enclosure shall not only resist blast pressures, but also provide penetration resistance (see Chapter 5) from façade fragments and structural debris. Safe haven structures shall have the strength and stability to resist blast pressures and potential impact from debris. The structure of the building above and below the safe haven shall be designed to avoid general collapse that would cause failure of the safe haven. Specific design requirements for elements of the enclosure and structure of the safe haven and structure of the building surrounding the safe haven shall satisfy the provisions of this Standard.

7.3.2 Applicable Loads and Performance. Safe haven load and performance criteria shall follow the same classification of loads and LOPs as for buildings. The safe haven shall have LOP IV, a high level of protection, regardless of the level of threat and protection of the surrounding building. Design pressure loading shall be based on the threat assessment, with consideration for the loads specified in NFPA 68. The portion of building surrounding the safe haven shall also have protection against general or progressive collapse and/or severe damage that would preclude the access to or use of the safe haven.

The walls, ceiling, and floor of the safe haven shall be hardened to provide LOP IV, a high level of protection, to both blast pressures and penetration due to fragment impact on the structural element. Openings, including windows, in any of these surfaces shall be minimized to avoid force concentrations. The ceiling of the safe haven may be an independent structure or a hardened section of the floor above. The walls of the safe haven shall extend from the floor to the ceiling of the safe haven and be appropriately connected thereto.

Although consideration may be given to the low probability of occurrence when combining extreme loads resulting from natural hazards with the effects of self-weight and live service design loads, no reduction of loads or increases in allowable resistance shall be considered for the design of the elements of the safe haven.

7.3.3 Resistance to Progressive Collapse.
If the entire structure containing a safe haven is designed to resist progressive collapse, it shall be permissible to use the walls of the safe haven as part of the load-resisting system for the building. If the rest of the structure is susceptible to progressive collapse, the safe haven shall not be structurally tied to the remaining structure unless it can be demonstrated that a partial collapse of some part of the structure will not lead to collapse of the safe haven.

7.3.4 Location within Building.
A standalone in-ground or aboveground safe haven shall be located in central proximity to intended users, typically in the center of a cluster of buildings. Standalone safe havens require careful site analysis with attention to standoff distances from points of potential threats. Because building materials and equipment during a blast event may become dislodged from tall buildings and fall upon the safe haven, safe havens shall be located away from a taller neighboring structure to minimize the potential impact.

Internal safe havens shall be located in an area of the building that can be quickly and easily accessed without exiting the building. Safe havens shall be selected according to the project risk assessment in order to provide the space required to accommodate the building population and shall be centrally located. Routes to the safe haven shall be easily accessible and well-marked.

Safe havens for new construction shall be located away from the building perimeter or shall be enclosed within an exterior hardened enclosure. The walls of such rooms shall not contain windows. For retrofit of existing structures, safe havens shall also be located away from the building perimeter unless no other feasible location exists. If a safe haven is selected at the building perimeter (retrofit only), the walls and windows shall be upgraded to resist the design basis threat. In multistory structures, safe havens shall be located on lower floors and not on the top floor unless the roof is appropriately designed to resist the blast effects associated with the design explosive.

7.3.5 Accessibility to Egress.
Safe havens shall be located near enclosed stairwells so that occupants of the safe haven can safely leave the building after the event. Exit routes from the safe haven shall be in a direction away from the origin of the explosive threat and away from roof and exterior glazing elements. If the safe haven is not attached to a staircase, then the safe haven shall have a minimum of two exit doors that open into two different and separated corridors. Emergency egress doors shall be designed according to the provisions of Section 8.3.6.

7.3.6 Fire Rating.
The fire rating used for the safe haven shall meet or exceed that used for the enclosed stairwells contained in the structure.

7.4 CONSENSUS STANDARDS AND OTHER REFERENCED DOCUMENTS

U.S. Department of Defense
Structures to Resist the Effects of Accidental Explosions, UFC 3-340-02, 2008

Chapter 8
EXTERIOR ENVELOPE

8.1 DESIGN INTENT

8.1.1 General. The exterior envelope of a building is the first line of defense in the event of an external explosion. The exterior envelope shall be designed in accordance with the design objectives stated in Section 3.2. This chapter discusses design procedures and requirements for elements of the building envelope that include fenestration, non-load-bearing exterior walls, roof systems, and hazard-mitigating retrofits.

8.1.2 Primary structural elements on the building exterior, such as load-bearing walls, exterior columns, and spandrel beams, shall be designed in accordance with Chapter 6. Materials and their method of analysis presented in Chapter 6 shall also be acceptable for the exterior envelope elements covered in this chapter.

8.2 DESIGN PROCEDURES

8.2.1 General. The exterior envelope shall be designed to resist the design blast pressure and impulse specified by the project-specific requirements. Exterior envelope elements shall be designed to satisfy the design requirements of this Standard.

Two design approaches are acceptable for nonstructural exterior envelope components: (1) the resistance-based design approach, and (2) the hazard-based design approach. The approach(es) adopted shall depend on the project-specific requirements and level of protection (LOP) established for the building.

For both approaches, exterior envelope elements shall meet the requirements given in Sections 3.3 and 3.4 for the design loads. The interaction between nonstructural components and the primary structural system shall be considered to prevent building collapse due to localized structural failure. Exterior envelope elements shall not be detailed to connect directly to vertical load-carrying elements, unless such elements are designed with greater resistance than the connecting elements.

For the resistance-based design approach, the design pressure and impulse shall be taken from the design basis threat at the worst-case location(s) along the defended perimeter of the building.

For the hazard-based approach, a constant value of pressure and impulse shall be adopted for the design of nonstructural exterior walls and glazing, which may be less than the actual blast loading produced by the design basis threat. For a medium level of protection (LOP III) or below, a triangular load with the design pressure and impulse shall equal 4 psi (0.028 MPa) and 28 psi-msec (0.193 MPa-ms), respectively, unless otherwise specified by the project-specific risk assessment. For other levels of protection, the design pressure and impulse shall be specified by the project-specific risk assessment. The hazard-based approach shall not be adopted for structural elements, such as columns, roofs, and exposed floor slabs, which are part of the exterior envelope.

8.2.2 Response Criteria. Standard exterior envelope systems shall meet the requirements of Chapter 3, including the response limits, as demonstrated by design calculations or blast testing. Glazing behavior shall meet the performance conditions given in Section 3.3.3. Door behavior shall meet the performance conditions given in Section 3.3.4. Nonstandard exterior envelope systems shall require peer review or blast testing per Sections 10.2 and 10.6, respectively.

8.2.3 Analytical Methods. Acceptable analytical methods and dynamic effects for the building envelope, with the exception of glazing, shall be in accordance with Section 6.2. Average strength factors and dynamic increase factors for building façade components shall be in accordance with Section 3.5. Acceptable analytical methods for glazing design shall be in accordance with Section 8.3.2.

8.2.4 Balanced Design. Balanced design entails an established hierarchy of component strength, where connections are designed for the maximum strength of the connecting components and elements supporting other elements are designed for the maximum strength of the supported elements. The exterior envelope shall incorporate a balanced design for all elements when the hazard-based approach is adopted or as required by the Authority Having Jurisdiction.

8.2.5 Flying Fragments. Frangible exterior envelope elements, such as canopies and louvers, shall be designed to minimize flying fragments for all levels of protection.

8.3 FENESTRATION

8.3.1 General. This section applies to fenestration, including blast-mitigating windows, curtain walls, skylights, and doors.

8.3.2 Blast-Mitigating Window Systems. Blast-mitigating window systems shall be designed using static or dynamic analysis in accordance with this section. It shall be permitted for window design to use static analysis in accordance with ASTM F2248 and ASTM E1300 for determining the glazing capacity for a medium level of protection (LOP III) or lower. Dynamic analysis shall be required for window systems meeting a high level of protection (LOP IV).

The glazing layup shall meet the glazing performance conditions established for the building in accordance with Section 3.3.3. Frame and mullions shall comply with the glazing system framing response limits given in Table 3-2 for SDOF analysis. A minimum frame bite shall be specified for window systems of 0.375 in. (10 mm) with or without structural sealant, unless otherwise specified by the project-specific requirements.

8.3.2.1 Glazing. Glass types shall be annealed, heat-strengthened, or fully tempered. Glazing shall consist of single laminated panes, insulating glass units (IGUs) with a laminated inner pane, glass-clad polycarbonate, or laminated plastics. Other glass or glazing types may be used if shown by blast testing and analysis to perform as required.

To meet the requirements of balanced design for window systems, glazing shall be designed to fail before any supporting elements or connections using either the hazard- or resistance-based approach.

For dynamic analysis, glass design for blast effects shall be based on a maximum probability of glass breakage of 500 breaks per 1,000, unless another statistical value can be justified. Post-failure performance of a window system shall be assessed using industry-standard computer programs or verified by blast testing in accordance with Section 10.6. Assessment and verification of window systems shall account for glazing failure, laminate failure, and bite failure.

For a high level of protection (LOP IV) requiring "no break" of the glazing, two load cases shall be dynamically analyzed to account for phasing: (1) positive phase only, and (2) positive and negative phases.

Additionally, for the hazard-based design approach, a minimum percentage of the building's glazing area as determined by project-specific requirements shall meet the specified performance condition for pressures and impulses resulting from the design basis threat located mid-width of the building at the defended perimeter. When not specified for a medium level of protection (LOP III), it shall be acceptable to use a minimum value equal to 90% of the total glazing area.

8.3.2.2 Frame and Mullion Design.
Structural frames and mullions of window systems shall consist of steel, aluminum, or a combination of the two. Other materials may be used if tested in accordance with Section 10.6.

Frames and mullions shall be designed to prevent two modes of failure: (1) detachment of elements from the supporting system, and (2) large deformation that leads to premature failure of the glazing. Response limits exceeding values in Section 3.4 shall be deemed acceptable if verified by blast testing or explicit dynamic finite element analysis with peer review per Chapter 10.

For dynamic analysis, frames and mullions shall resist the glazing capacity based on a minimum probability of glass breakage equal to 500 breaks per 1,000 where balanced design is required, unless a lower statistical value can be justified.

8.3.2.3 Connections and Anchorage.
The ultimate strength of the connections and anchorage system shall exceed that of the frame and mullions by a minimum safety factor of 1.5.

For dynamic analysis, the connections and anchorage system shall resist the glazing capacity based on a minimum probability of glass breakage equal to 500 breaks per 1,000 where balanced design is required, unless a lower statistical value can be justified. In addition, the connections and anchorage system shall be designed to resist the smaller of the two static loads: the ultimate flexural resistance of the frame components, or the peak dynamic reactions from the frame.

8.3.2.4 Rebound Loads.
Load reversals from dynamic rebound shall be considered for medium or high levels of protection (LOP III or IV). For low and very low levels of protection, rebound shall be considered only if required by project-specific requirements. Rebound loads shall equal or be less than the peak loads from the positive phase of an airblast.

8.3.3 Curtain Wall Systems.
The glazed portion of curtain wall systems shall meet the requirements for window systems given in Section 8.3.2. Curtain walls, supports, or connections that are not part of a standard building envelope system shall require peer review or blast testing in accordance with Sections 10.2 and 10.6.

8.3.4 Skylights.
Skylights shall meet the requirements for window systems given in Section 8.3.2. Skylight glazing shall remain in the frame for all levels of protection to reduce glass hazard. Laminated glazing used in conjunction with a hazard-mitigating system, such as an interior catcher system, shall not be required to remain in the frame.

8.3.5 Operable Windows.
Blast-mitigating operable window systems shall be permitted if the requirements in Section 8.3 are met.

8.3.6 Doors.
When required by the project-specific risk assessment, blast-resistant doors shall meet the requirements of Section 3.3.4. Blast testing for doors shall comply with Section 10.6. The ultimate strength of the connections and anchorage system shall exceed that of the door by a minimum safety factor of 1.5.

8.3.6.1 Non-Blast-Resistant Doors.
Non-blast-resistant doors on the exterior envelope shall be selected and installed to minimize fragmentation hazard. When possible, non-blast-resistant exterior doors shall be designed to open outward with a bearing surface for the positive phase of the blast load. Provisions for non-blast-resistant doors shall be determined by project-specific requirements.

8.3.6.2 Public-Access Glass Doors.
Public-access glass doors shall consist of a tempered or laminated layup in accordance with ANSI Z97.1. Glass in doors shall comply with the requirements for blast-mitigating glazing in Section 8.3.2. Door frames shall be supported by full-length jambs and regularly spaced hinges. Door jambs shall be designed for equivalent peak static or dynamic reactions generated by the door and frame response. Reactions from door response shall be considered in overall storefront glazing applications.

8.3.6.3 Oversized and Rollup Flexible Metal Doors.
It shall be permitted to use oversized and rollup flexible metal doors that comply with performance conditions in Section 3.3.4 and meet blast testing requirements given in Section 10.6.

8.3.6.4 Other Doors.
Sections of egress, equipment, maintenance, delivery, and utility doors shall be designed to remain elastic for doors required to remain operational. Commercial or off-the-shelf blast-mitigating door and frame systems shall comply with the requirements in this section. Non-blast-resistant commercial door installations shall be retrofitted to comply with this section.

8.4 NON-LOAD-BEARING EXTERIOR WALLS

8.4.1 General.
This section applies primarily to non-load-bearing exterior walls. Requirements for load-bearing exterior walls are provided in Chapter 6. When conflicts arise between Chapters 6 and 8, the more stringent requirement shall apply.

8.4.1.1 Exterior walls shall withstand loads acting directly on their surface, in addition to reactions from adjacent glazing systems and blast-resistant doors. Nonductile modes of failure, such as shear, punching shear, connection, or anchorage failure shall be avoided. Exterior walls shall be designed to prevent localized failure in areas near openings. Rebound shall be considered for exterior elements if required by the project-specific requirements.

8.4.1.2 A balanced design shall be provided between two-way walls and supporting columns, where structural columns have greater resistance to airblast effects than walls to reduce the risk of progressive collapse.

8.4.1.3 Envelope walls with forced entry and ballistic resistance requirements shall also mitigate blast in accordance to project-specific requirements.

8.4.1.4 For the hazard-based design approach, the surviving wall area meeting the response limits for pressures and impulses resulting from the design basis threat located mid-width of the building at the defended perimeter shall remain within the percentage specified by the project-specific requirements.

8.4.2 Cast-in-Place, Precast, and Tilt-Up Concrete Walls. Cast-in-place reinforced concrete walls, precast panels, and tilt-up construction shall be designed to provide protection against airblast effects. The response limits for concrete slabs in Section 3.4 shall apply to concrete walls for SDOF analysis of far-range loads. When using other methods of analysis, such as explicit dynamic finite element analysis, or evaluating near-range loads, requirements for structural elements given in Chapter 6 shall apply. Where wall reinforcement is interrupted due to openings, reinforcement shall be designed to prevent localized failure. Rebound shall be considered for all concrete wall design.

8.4.3 Pretensioned and Posttensioned Concrete Wall Panels. Tendons of pretensioned and posttensioned panels shall be straight and fully grouted. Sections and anchorages shall be detailed to allow full development of the steel tendon.

8.4.4 Masonry Walls

8.4.4.1 Reinforced concrete masonry units (CMUs) shall be fully grouted to allow full development of the reinforcing steel in flexure to resist the applied blast load, unless otherwise specified by the project-specific requirements. Reinforced CMU walls shall be designed as either one- or two-way systems using cell and joint reinforcement. Shear resistance shall be provided by the CMU in combination with steel reinforcement.

When permitted by project-specific requirements, partially grouted CMU walls shall meet the levels of protection and performance conditions in Section 3.3.2, as demonstrated by blast testing in compliance with Section 10.6. Partially grouted CMU walls shall have all reinforced cells grouted.

8.4.4.2 Unreinforced masonry shall not be permitted for new construction. For existing unreinforced masonry walls, wall stability shall be assessed. The stability assessment shall assume that the wall is simply supported if flexible supports are provided. It shall be permitted to assume arching action of the masonry wall if rigid supports are provided in accordance with UFC 3-340-02.

8.4.4.3 Alternatively, debris catch systems shall be permitted to minimize the debris impact hazard. The effectiveness of the debris catch systems shall be determined through blast testing or advanced finite element analysis.

8.4.5 Steel Wall Systems. Steel elements shall be designed to develop their full design capacity without connection or anchorage failure. Bracing shall be provided for steel elements designed in accordance with UFC 3-340-02. Minimum thickness requirements shall be met to prevent local buckling. Unsymmetrical elements shall not be used unless restraints are shown to prevent twisting. Nested elements shall not be used as composite elements unless shear continuity is provided to ensure composite behavior. Steel sections shall be designed with blast-mitigating panels or a non-blast-mitigating façade with a catch system located on the interior face. Plate edges and supporting elements shall be designed to resist tension membrane forces, if membrane action of the element is assumed.

It shall be permitted to use other types of steel wall construction that meet the requirements in this chapter and the blast testing requirements in Section 10.6.

8.4.6 Other Wall Systems. Other wall systems shall meet the general requirements in Section 8.4.1.

8.5 ROOF SYSTEMS

8.5.1 General

8.5.1.1 The resistance-based approach shall be adopted for the design of roof systems. Roof elements shall resist incident pressures and impulses produced by an exterior explosion. Acceptable methods for calculating incident blast loads are given in Chapter 4.

8.5.1.2 Roof systems shall be designed to resist flexure and shear loads. Component design shall force ductile failure modes to precede nonductile failure modes. Roof framing shall be designed as primary structural elements in accordance with Section 6.2, as its failure could lead to progressive collapse. Load reversals due to dynamic rebound shall be considered for roof design.

8.5.1.3 Membrane action shall be considered only if the roof slab or metal deck has sufficient continuity and detailing to allow transfer of in-plane tension forces between bays, and supporting elements have sufficient capacity to resist these forces and their reactions.

8.5.2 Flat Slabs. The requirements for reinforced concrete in Section 3.4 shall apply to concrete slabs. Reinforced concrete slabs shall be detailed to allow full development of the section in flexure, in accordance with Section 9.2.

8.5.3 Metal Deck. Cold-formed metal deck shall conform to requirements in Section 3.4.

8.5.4 Composite Construction. Composite steel-concrete construction for roof systems shall be detailed in accordance with Section 9.4. Roof systems designed for composite action shall account for load reversals due to concrete crushing during the positive phase of an airblast.

8.5.5 Steel Joists and Joist Girders. Steel joists and joist girders shall be designed in accordance with Chapter 6. In addition, they shall develop fully plastic sections in flexure without buckling of chord or bridging elements. Component design shall force ductile failure modes to precede nonductile failure modes.

8.5.6 Other Roof Systems. Other roof systems shall be permitted that meet the general requirements in Section 8.5.

8.6 OTHER EXTERIOR ENVELOPE ELEMENTS

Other elements of the exterior envelope include exposed floor systems, cantilever beams, and spandrel beams. Exposed floor systems shall be designed in accordance with the resistance-based design approach. Primary structural elements on the exterior envelope shall be designed in accordance with Chapter 6. Secondary structural elements, such as canopies and appurtenance elements, shall be designed to meet the requirement in Section 8.2.5 unless otherwise required by project-specific requirements.

8.7 HAZARD-MITIGATING RETROFITS

8.7.1 General. This section discusses hazard-mitigating retrofits used to capture fragments or strengthen envelope components.

Window retrofits improve the performance condition of existing glazing systems. Exterior wall retrofits improve the structural response and reduce the fragmentation hazard of existing systems.

It shall be demonstrated that a hazard-mitigating retrofit meets the levels of protection and performance conditions given in Sections 3.3 and 3.4 by calculation or by blast testing in compliance with Section 10.6.

Additional hazard-mitigating materials or systems not described in this Standard shall be permitted if the requirements of this section are met and verified by blast testing.

8.7.2 Security Window Films. It shall be permitted to retrofit existing windows with security window films adhered to the interior surface of the glass. Four types of security window film applications shall be acceptable: daylight installation, edge-to-edge installation, wet-glazed installation, or mechanical attachment systems.

Security window films shall have a minimum thickness of 0.004 in. (0.10 mm). The membrane strength of the film shall be considered for edge-to-edge installation, wet-glazed installation, and mechanical attachment systems. The minimum overlap shall be 0.25 in. (6 mm) between the film and the silicone bead, and 0.25 in. (6 mm) between the frame and the silicone bead for wet-glazed installations.

8.7.3 Blast Curtains. Two types of blast curtain systems shall be permitted for the retrofit of existing glazing systems: (1) a taut system with rigidly held synthetic fabric, or (2) a dynamic tension system with initial slack. Blast curtains shall be secured on at least two opposite sides, unless demonstrated by blast testing to meet the performance conditions in Section 3.3.3.

8.7.4 Catch Bar Systems. Catch bar systems shall be designed to catch the glazing after it fails. Daylight films, laminated glass, or special laminate interlayers shall be used to hold the glazing in one piece after glass failure. The catch bars shall be designed to resist the maximum capacity of the glazing. The system shall be designed to prevent slicing of the failed glazing into two or more pieces, which shall be verified by calculation or blast testing in compliance with Section 10.6. Catch bar systems shall not be used with insulating glass units (IGUs) unless all panes are laminated. Design for dynamic rebound shall not be required for catch bar systems. Response limits in Section 3.4 are not applicable to catch bar systems.

8.7.5 Secondary Window System. A secondary blast-mitigating window system installed on the inside face of an existing window system shall comply with the requirements in Section 8.3.2.

8.7.6 Window Replacement. Replacement of an existing window with a blast-mitigating window shall comply with the requirements in Section 8.3.2.

8.7.7 Geotextile Fabrics. It shall be permitted to use geotextile fabrics to act as a permanent blast curtain installed on the inside face of a non-blast-mitigating concrete or masonry wall. The geotextile fabric shall be designed for tension membrane response. The connections shall transfer loads from the fabric into the floor system without fabric disengagement or tears from stress concentrations. The performance of a specified geotextile fabric and its connections shall be verified by blast testing in compliance with Section 10.6.

8.7.8 Fiber-Reinforced Polymers. It shall be permitted to retrofit an existing concrete or masonry wall or slab system with fiber-reinforced polymers (FRP) to improve its structural response. FRP wall retrofits shall be designed and detailed in accordance with Section 9.6. The building structure shall be re-evaluated to ensure that global load redistribution through the lateral system due to the wall retrofit does not lead to structural failure of primary elements for all load combinations.

8.7.9 Secondary Wall System. A secondary blast-mitigating wall system installed on the inside face of an existing wall system shall comply with the requirements in Section 8.4. It shall be permitted to install a secondary wall system as a retrofit behind a non-blast-resistant door in a vestibule configuration.

8.7.10 Other Retrofits. Other blast-mitigating retrofits shall be permitted if the requirements in Section 8.7 are met and verified by blast testing in compliance with Section 10.6.

8.8 AMPLIFICATION AND REDUCTION OF BLAST LOADS

8.8.1 Building Shape and Site. Amplification or reduction of blast loads on the exterior envelope due to the building shape and site shall be considered when the resistance-based design approach is adopted.

8.8.2 Venting. Amplification of blast loads due to internal threats shall be considered in accordance with Chapter 4. Explosive forces and gases shall be vented from interior spaces to the building exterior in accordance with the project-specific requirements. Venting shall be provided by openings, vent shafts, blowout panels, non-blast-resistant doors, or directional window systems that provide venting for internal explosions and protection from exterior explosions. Design for venting shall be in accordance with UFC 3-340-02 or NFPA 68.

8.9 CONSENSUS STANDARDS AND OTHER REFERENCED DOCUMENTS

ANSI
American National Standards Institute
25 West 43rd Street, 4th Floor
New York, NY 10036
Safety Glazing Materials Used in Buildings—Safety Performance Specifications and Methods of Test, ANSI Z97.1, 2004.

ASTM International
100 Barr Harbor Drive
P.O. Box C700
West Conshohocken, PA 19428-2959
Standard Practice for Specifying an Equivalent 3-Second Duration Design Loading for Blast Resistant Glazing Fabricated with Laminated Glass, ASTM F2248, 2009.
Standard Practice for Determining Load Resistance of Glass in Buildings, ASTM E1300, 2007.

NFPA
National Fire Protection Association
1 Batterymarch Park
Quincy, MA 02169-7471
Standard on Explosion Protection by Deflagration Venting, NFPA 68, 2007.

U.S. Department of Defense Explosives Safety Board (DoD-ESB)
Room 856C Hoffman Building #1
2461 Eisenhower Avenue
Alexandria, VA 22331-0600
Structures to Resist the Effects of Accidental Explosions, UFC 3-340-02, 2008.

Chapter 9
MATERIALS DETAILING

9.1 GENERAL

9.1.1 Scope. Materials, design, detailing, and construction of primary and secondary elements intended to resist the effects of blast shall comply with the requirements of the references listed in each section of this chapter, except as specifically modified. The provisions of this chapter are intended primarily for new construction. For evaluation of existing construction, refer to UFC 3-340-02.

9.1.2 Structural Interaction. Individual element and global structural performance shall comply with Sections 9.1.2.1 and 9.1.2.2.

9.1.2.1 Primary and Secondary Elements. All elements that interact structurally with a primary element shall be detailed to support the intended performance of that element for the specified LOP, regardless of whether they are otherwise identified as secondary elements.

9.1.2.2 Predominance of Global Structural Performance. Element detailing shall not violate the provisions of Section 3.3.1 of this Standard.

9.1.3 Materials. It shall be permissible to use materials other than specified in this chapter if such material is shown by testing or rational analysis to satisfy the performance requirements of this Standard.

9.1.4 Detailing. It shall be permissible to use detailing other than specified in this chapter if shown by testing or rational analysis to satisfy the performance requirements of this Standard.

9.1.5 Controlling Loads. The provisions in this chapter shall be satisfied for any elements of the structure so designated during the design process as potentially being subjected to the effects of accidental or malicious explosions. In the event that load combinations not containing the blast effects indicate larger demands than load combinations including blast, the element shall be detailed to meet both the larger strength demands and satisfy the provisions of this chapter.

9.1.6 Achieving Design Intent. This chapter provides essential, but not necessarily sufficient, provisions for detailing to resist the effects of blast loading. The registered design professional must determine whether additional detailing is required to meet the specified LOP.

9.1.7 Use of Reference Documents. Elements shall be detailed according to the complete provisions of the reference documents in Section 9.8, unless otherwise noted.

9.1.8 Special Inspection. Special inspection, according to standards for seismic inspections acceptable to the Authority Having Jurisdiction, is mandated for all blast-resistant construction.

9.2 CONCRETE

9.2.1 Scope. Requirements are provided for columns, beams, slabs, and walls subject to blast loads. Precast concrete and tilt-up construction are not addressed in this chapter of the Standard.

9.2.2 General Reinforced Concrete Detailing Requirements
9.2.2.1 Specified compressive strength of concrete shall be not less than 3,000 psi (20.7 MPa).

9.2.2.2 All concrete shall be normal-weight unless demonstrated through test or rational analysis that light-weight concrete can satisfy the performance requirements of this standard.

9.2.2.3 All reinforcement shall comply with ASTM A706 or ASTM A615 with limitations as specified in Chapter 21 of ACI 318-08 and exceptions as noted below.

9.2.2.4 Elements containing lapped, mechanical, or welded spliced reinforcing steel shall not be relied upon to develop full tension membrane action.

9.2.2.5 All lap splices shall be at least Class A per ACI 318-08 and enclosed in special transverse reinforcement per Sections 21.6 of ACI 318-08. No more than 50% of bars at a cross section may be spliced at a given location. All lap splices shall be separated along the axis of an element by at least twice the lap length.

9.2.2.6 Mechanical splices shall develop the dynamic design stress of the bar under conditions generated by blast loads (including strain rate) and shall be limited to zones that are designed to remain elastic during blast response unless ASTM A706 reinforcing steel is used. No more than 50% of bars at a cross section may be spliced at a given location. All mechanical splices shall be separated along the axis of an element by at least the lap development length.

9.2.2.7 Welded splices shall be limited to zones that remain elastic during blast response, unless the welded splices can be shown to develop the dynamic design stress of the bar under conditions generated by blast loads (including strain rate), per the provisions of Chapter 10. No more than 50% of bars at a cross section may be spliced at a given location. Welded splices shall be separated along the axis of an element by at least the lap development length. Welded splices shall only be made in ASTM A706 reinforcement.

9.2.2.8 Columns are defined as elements for which the factored axial compressive force exceeds $0.10A_g f'_c$. Beams are defined as elements for which the factored axial compressive force does not exceed $0.10A_g f'_c$.

9.2.3 Columns
9.2.3.1 The limits on longitudinal reinforcement ratios are given in Section 21.6.3.1 of ACI 318-08.

9.2.3.2 Transverse reinforcement per Sections 21.6.4.1, 21.6.4.2, and 21.6.4.3 of ACI 318-08 shall be provided over the clear height of the column.

9.2.3.3 The design shear force for checks of diagonal tension and compression shall be determined per Section 21.6.5.1 of ACI 318-08, except that M_{pr} shall be computed using $1.25f_{ds}$ in lieu of $1.25f_y$, where f_{ds} is the product of the specified yield stress of the reinforcement, the average strength factor (ASF) and the dynamic increase factor (DIF) from Chapter 3. Transverse reinforcement for shear shall be proportioned assuming that the concrete contribution, V_c, is computed per Sections 11.2.1.2 and 11.2.2.3 of ACI 318-08. When axial loads due to blast or other loading result in net tension in reinforced concrete elements, then the shear capacity of the concrete, V_c, shall be neglected.

9.2.3.4 The design shear force for direct shear shall be computed per Chapter 6 of this standard. The direct design shear strength shall be based on a combination of the strengths of the concrete and inclined bars. If the design support rotation, θ, is less than or equal to 2 deg, it shall be permissible to compute the shear strength as the sum of the nominal axial strength of inclined bars and $0.18f'_{cd}bd$. The nominal axial strength of diagonal bars shall be calculated as:

$$V_s = A_d f_{ds} \sin\alpha \qquad (9\text{-}1)$$

where A_d is the total area of diagonal bars; f_{ds} is the product of the specified yield stress of the reinforcement, the ASF and the DIF from Chapter 3; and α is the angle formed by the plane of the diagonal reinforcement and the longitudinal reinforcement. If the design support rotation exceeds 2 deg, the shear strength shall be computed as nominal axial strength of inclined bars alone. When axial loads due to blast or other loading result in net tension in reinforced concrete elements, then the shear capacity of the concrete, V_c, shall be neglected.

9.2.4 Beams

9.2.4.1 The clear span of a beam shall not be less than four times its effective depth to use the requirements of this section. Other beams are beyond the scope of this chapter.

9.2.4.2 Longitudinal reinforcement shall comply with Sections 21.5.2.1 and 21.5.2.2 of ACI 318-08.

9.2.4.3 Transverse reinforcement per Section 21.5.3.2 of ACI 318-08 shall be provided over the clear span of the beam. Stirrups or hoops shall be made up of one piece of reinforcement with 135-deg hooks per Detail A as shown in Fig. R21.5.3 of ACI 318-08.

9.2.4.4 The design shear force for diagonal tension and compression shall be determined per Section 21.5.4.1 of ACI 318-08. It shall be permissible to proportion the transverse reinforcement for shear assuming that the concrete contribution, V_c, is computed per Sections 11.2.1.1 and 11.2.2.3 of ACI 318-08. When axial loads due to blast or other loading result in net tension in reinforced concrete elements, then the shear capacity of the concrete, V_c, shall be neglected.

9.2.4.5 The design shear force for direct shear shall be computed per Chapter 6 of this Standard. The direct design shear strength shall be based on the nominal axial strength of inclined bars. If the design support rotation, θ, is less than or equal to 2 deg, the shear strength can be computed as the sum of the nominal axial strength of inclined bars and $0.18f'_{cd}bd$. The nominal axial strength of diagonal bars shall be calculated as:

$$V_s = A_d f_{ds} \sin\beta \qquad (9\text{-}2)$$

where A_d is the total area of diagonal bars; f_{ds} is the product of the specified yield stress of the reinforcement, the ASF and the DIF from Chapter 3; and β is the angle formed by the plane of the diagonal reinforcement and the longitudinal reinforcement. If the design support rotation exceeds 2 deg, the shear strength shall be computed as nominal axial strength of inclined bars alone.

9.2.4.6 Open stirrups, either single- or double-leg, are not permitted.

9.2.5 Beam-Column Joints

9.2.5.1 Forces in longitudinal beam reinforcement at the joint face shall be determined by assuming that the stress in the flexural reinforcement is the dynamic design stress (as determined by the methods in Chapter 6).

9.2.5.2 Longitudinal beam reinforcement details shall comply with Sections 21.7.2.2 and 21.7.2.3 of ACI 318-08. Development lengths for bars in tension shall comply with Section 21.7.5 of ACI 318-08.

9.2.5.3 Transverse hoop reinforcement in columns shall be provided within the joint without any reduction based on the framing around the joint.

9.2.5.4 Shear strength shall be computed per Section 21.7.4 of ACI 318-08. When axial loads due to blast or other loading result in net tension in reinforced concrete elements, then the shear capacity of the concrete, V_c, shall be neglected.

9.2.6 Slabs

9.2.6.1 The minimum flexural reinforcement ratio in a slab shall be 0.0018.

9.2.6.2 For slabs with total thickness not exceeding 6 in. (150 mm), flexural reinforcement may be placed in a single layer at mid-depth. Otherwise, symmetrical top and bottom flexural reinforcement shall be provided.

9.2.6.3 Bottom reinforcement in a slab shall not be curtailed and shall be continuous over interior supports (splices are not permitted). Bottom reinforcement in a slab shall be fully developed at the face of an exterior support.

9.2.6.4 In two-way flat plate and flat slab floor systems, column strip bottom reinforcement shall be spaced uniformly over the column strip, shall not be curtailed, and shall be continuous over interior supports (splices are not permitted) and fully developed at the interior face of exterior supports.

9.2.6.5 The design shear force for diagonal tension shall be determined per Section 21.5.4.1 of ACI 318-08. Transverse reinforcement for shear can be proportioned assuming that the concrete contribution, V_c, is computed per Sections 11.2.1.1 and 11.2.2.3 of ACI 318-08. When axial loads due to blast loads result in net tension in reinforced concrete elements, then the shear capacity of the concrete, V_c, shall be neglected.

9.2.6.6 The design shear force for direct shear shall be computed per Chapter 6 of this Standard. The direct design shear strength shall be based on the nominal axial strength of inclined bars. If the design support rotation, θ, is less than or equal to 2 deg and the slab is not subjected to net tension, the shear strength can be computed as the sum of the nominal axial strength of inclined bars and $0.18f'_{cd}bd$. The nominal axial strength of diagonal bars shall be calculated as:

$$V_s = A_d f_{ds} \sin\beta \qquad (9\text{-}3)$$

where A_d is the total area of diagonal bars; f_{ds} is the product of the specified yield stress of the reinforcement, the ASF and the

DIF from Chapter 3; and β is the angle formed by the plane of the diagonal reinforcement and the longitudinal reinforcement. If the design support rotation exceeds 2 deg, the shear strength shall be computed as nominal axial strength of inclined bars alone.

9.2.6.7 Slabs functioning as tension membranes shall have reinforcement that is fully developed in those components anchoring the membrane.

9.2.7 Walls

9.2.7.1 The minimum horizontal and vertical reinforcement ratios in the web of a wall resisting the in-plane blast loads shall be 0.0025. Web reinforcement shall be fully developed at wall boundaries. All reinforcement splices shall be at least Class A per ACI 318-08. Horizontal and vertical reinforcement splices shall be staggered.

9.2.7.2 The minimum horizontal and vertical reinforcement ratios in the web of a wall resisting out-of-plane blast loads shall be 0.0018. Web reinforcement shall be fully developed at wall boundaries. All reinforcement splices shall be at least Class A per ACI 318-08. Horizontal and vertical reinforcement splices shall be staggered.

9.2.7.3 The in-plane shear strength of a structural wall shall be calculated per Section 21.9 of ACI 318-08. The out-of-plane shear strength shall be computed assuming that the wall is a slab.

9.2.8 Tension Ties

9.2.8.1 Tension ties shall be designed and detailed assuming that only reinforcement resists tensile loads.

9.2.8.2 Reinforcement area shall be computed using the specified minimum yield strength without increase for dynamic loading.

9.2.8.3 Reinforcement can be spliced by laps, mechanical devices, or by welding. Lap splices shall be Class B per ACI 318-08 and staggered. Mechanical splices shall be Type 2, as defined by ACI 318-08, and additionally shall meet the requirements of Section 9.2.2.6 of this Standard. Welded splices shall comply with Section 9.2.2.7 of this Standard.

9.3 STRUCTURAL STEEL

9.3.1 Scope. Requirements for structural steel elements are as defined in documents referenced in Section 9.8. Cold-formed, stainless, and other steel materials are not addressed in this section of the Standard.

9.3.2 General Structural Steel Requirements

9.3.2.1 Materials: steel materials shall satisfy Sections 6.1 and 6.3 of ANSI/AISC 341.

9.3.2.2 Welds: welded joints shall satisfy Section 7.3 of ANSI/AISC 341.

9.3.2.3 Bolts: bolted joints shall satisfy Section 7.2 of ANSI/AISC 341.

9.3.2.4 Connections: connections shall satisfy Section 7.1 of ANSI/AISC 341, with the ductile limit state checked for the loads determined by analysis. Connections need not exceed required element strength.

9.3.2.5 Element Slenderness Limits

9.3.2.5.1 Flanges and webs in LOP I and II construction in which inelastic deformations are expected shall comply with the limits on λ_p for flexure in ANSI/AISC 360.

9.3.2.5.2 Flanges and webs in LOP III construction shall comply with the limits on λ_p for flexure in ANSI/AISC 360, even if elastic.

9.3.2.5.3 Flanges and webs in LOP IV construction shall comply with the limits on λ_{ps} for flexure in ANSI/AISC 341.

9.3.2.5.4 Noncompact sections may only be used if shown by nonlinear, dynamic finite element analysis or testing to meet the specified LOP.

9.3.2.6 It shall be permissible to multiply F_y and F_u by R_y and R_t of Section 6 of ANSI/AISC 341, respectively, in lieu of the average strength factor (ASF) from Chapter 3 of this Standard.

9.4 STEEL/CONCRETE COMPOSITE

9.4.1 Scope. This section provides requirements for concrete slabs on metal deck. These provisions are minimum requirements and are not necessarily sufficient, particularly for instances of interior and close-in blast. Composite beams and columns are not addressed in this edition of the standard.

9.4.2 Concrete Slab on Metal Deck

9.4.2.1 Deck panels shall be anchored to supporting elements including perimeter elements and walls by welding or mechanical fastening in accordance with SDI No. 31.

9.4.2.2 Design of the concrete slabs on metal deck shall be in accordance with ACI 318-08 requirements for strength, serviceability, and integrity.

9.4.2.3 Floor slabs shall have a minimum thickness above the metal deck of 3 in. (76 mm) of normal-weight concrete. It shall be permissible to use light-weight concrete if shown by analysis or testing to satisfy the performance requirements of the specified LOP.

9.4.2.4 Floor slabs shall be continuously reinforced with bars in each direction or equivalent welded wire reinforcement. Where wire reinforcement is used, lap splices shall cover at least two mesh squares, and meet or exceed the requirements of ACI 318-08.

9.4.2.5 Where reinforcing bars are utilized, the top layer shall be below the top of the headed stud and at least one bar shall be placed in contact with headed studs parallel to each beam.

9.4.2.6 Reinforcement splices shall be class B per ACI 318-08 and shall be staggered at least 48 in. (1.2 m), with no more than one-third of the reinforcement spliced at any location.

9.4.2.7 Reinforcement splices are not allowed within 6 ft (1.8 m) from a column centerline or face of a supporting beam.

9.4.2.8 The concrete slab shall be designed to provide sufficient (at least 50%) composite action with the supporting steel element to prevent slab separation.

9.4.2.9 Horizontal shear connections shall be achieved using steel headed studs as prescribed in ANSI/AISC 360 for complete shear connection.

9.4.2.10 A minimum of one headed stud every 12 in. (305 mm) shall be provided.

9.4.2.11 Studs shall extend up to within 0.5 in. (13 mm) of the top of slab.

9.5 MASONRY

9.5.1 Scope. Requirements for masonry.

9.5.2 General Design Requirements

9.5.2.1 Strength design shall be used to design masonry components for blast loadings.

9.5.2.2 The registered design professional shall specify block type and grouting requirements, as well as horizontal and vertical bar size, bar surface treatment, and bar material at all locations in the blast-resistant masonry structural elements.

9.5.3 General Material Requirements

9.5.3.1 All reinforcement in new construction shall comply with ASTM A706 for LOP III and IV and ASTM A706 or ASTM A615 for LOP I and II.

9.5.3.2 Maximum masonry strength, f'_m, shall not exceed 6,000 psi (41.4 Mpa) for all LOP in new construction

9.5.3.3 Minimum masonry strength, f'_m, shall not be less then 1,500 psi (10.3 Mpa) for all LOP in new construction

9.5.3.4 The registered design professional shall determine the material properties of the existing masonry structural elements to ensure proper design to achieve the LOP required for the existing masonry structure.

9.5.4 General Detailing Requirements

9.5.4.1 Unreinforced masonry is not permitted.

9.5.4.2 All concrete masonry units (CMUs) shall be fully grouted for LOP III and LOP IV.

9.5.4.3 All lap splices shall be tension lap splices as specified in ACI 530/530.1-08.

9.5.4.4 Use of mechanical splices shall be limited to LOP I and LOP II, and to zones that remain elastic during blast response, unless the mechanical splices can be shown to develop the dynamic design stress of the bar under conditions generated by blast loads, per the provisions of Chapter 10.

9.5.4.5 Use of welded splices shall be limited to LOP I and LOP II, and to zones that remain elastic during blast response, unless the welded splices can be shown to develop the dynamic design stress of the bar under conditions generated by blast loads, per the provisions of Chapter 10. All welded splices shall meet or exceed the specifications in ACI 530/530.1-08.

9.5.4.6 The registered design professional shall direct the detailing on the design documents such that all requirements of the blast-resistant design rework and renovation on an existing structure are met. The design engineer shall verify that the design documents clearly detail any required repairs of the existing structure, modifications to the existing structure, or further work to the existing structure in order to achieve the LOP required of the modified existing structure.

9.5.4.7 Shear strength of masonry walls shall exceed shear demand due to development of ultimate flexural capacity as specifies in ACI 530/530.1-08 with appropriate support conditions.

9.5.5 Walls

9.5.5.1 Vertical Reinforcement

9.5.5.1.1 All elements constructed to LOP I and II shall have continuous vertical reinforcing bars in a fully grouted cell column with spacing not to exceed 48 in. (1.2 m) on center.

9.5.5.1.2 All elements constructed to LOP III and IV shall have at least one continuous vertical reinforcing bar per block in a fully grouted cell column with spacing not to exceed 16 in. (406 mm) on center.

9.5.5.1.3 At least one vertical bar shall be placed in a fully grouted cell column at all inside or outside corners, alongside any openings, and in the cell column on either side of control joints if used in the structure.

9.5.5.1.4 All elements constructed to LOP III and IV shall have minimum vertical reinforcement ratio not less than 0.0025.

9.5.5.2 Horizontal Reinforcement

9.5.5.2.1 A bond beam course with two horizontal reinforcing bars shall be placed all around the cap course of the wall; bar shall be spliced and course grouted per requirements of ACI 530/530.1-08.

9.5.5.2.2 Shear reinforcement of masonry shall be per requirements of ACI 530/530.1-08 and shall be placed so that every 45-deg line extending from mid depth ($dc/2$) of a wall to the tension bars crosses at least one line of shear reinforcement.

9.5.5.2.3 The design shear force for direct shear shall be computed per Chapter 6 of this standard. The direct design shear strength shall be based on the capacity of the steel longitudinal bars as specified in ACI 530/530.1-08.

9.5.5.3 All blast-resistant construction shall have a bond beam course with a minimum of two horizontal reinforcing bars that begins below the bearing of the diaphragm and shall be placed at any horizontal diaphragm or floor system supported by the wall; bars shall be spliced and course grouted per requirements of ACI 530/530.1-08. The cells of any and all courses above the bond beam to the top of the diaphragm shall be fully grouted.

9.5.5.3.1 A bond beam course with two horizontal reinforcing bars shall be placed at all lintel locations; bar shall be either hooked around vertical reinforcement on either side of opening or continuous along the wall with course grouted per requirements of ACI 530.1.

9.5.5.4 Control Joints

9.5.5.4.1 Control joints for all LOPs shall be designed with one vertical bar in a fully grouted cell column on either side of the control joint as a minimum. Code-approved control joint block and expansion material systems shall be used as necessary in all blast-resistant masonry systems.

9.5.5.4.2 The registered design professional shall determine the appropriate locations for control joints, to the extent that their locations impact blast resistance.

9.5.5.4.3 The registered design professional shall specify continuity and/or end conditions of horizontal bars and bond beams at control joints.

9.6 FIBER REINFORCED POLYMER (FRP) COMPOSITE MATERIALS

9.6.1 Scope. This section provides requirements for the following FRP systems:

- FRP composite hybrid
  - FRP strengthened steel reinforced concrete (RC) beams and slabs
  - FRP strengthened masonry
  - FRP column confinement
- Thick laminate FRP plates and shells

9.6.2 Strength Increase Factors

9.6.2.1 The average strength factor (ASF), as defined in Chapter 3 of this Standard, shall be taken as 1.0 unless otherwise determined by rational statistical analysis of material data.

9.6.2.2 The dynamic increase factor (DIF), as defined in Chapter 3 of this Standard, shall be taken as 1.0 unless otherwise determined by rational statistical analysis of material data.

9.6.3 General
9.6.3.1 FRP Delamination Due to Stress Wave Propagation.
The unreinforced resin matrix shall exhibit a static tensile strength of at least 10,000 psi (69 MPa).

9.6.4 FRP Strengthened Reinforced Concrete Beams and Slabs
9.6.4.1 FRP strengthening of steel reinforced concrete (RC) beams and slabs shall not be implemented for compressive reinforcement, in accordance with ACI 440.2R.

9.6.4.2 All FRP strengthening reinforcement shall comply with ACI 440.2R and specifically the reinforcement detailing requirements (Section 12) and drawing requirements (Section 13) contained therein.

9.6.4.3 FRP shall be attached to concrete elements using mechanical anchorage. Bonded FRP shall not be considered as providing blast resistance unless demonstrated by testing.

9.6.5 FRP Strengthened Masonry Walls
9.6.5.1 FRP strengthening of masonry walls shall not be implemented for reinforcement of the compressive face in accordance with ACI 440.2R.

9.6.6 FRP Concrete Column Confinement Requirements and Limitations
9.6.6.1 This section only addresses confinement of reinforced concrete columns to increase ductility.

9.6.6.2 All FRP column reinforcement shall comply with ACI 440.2R.

9.6.6.3 FRP reinforcement designed to provide ductility enhancement shall not be considered to provide additional capacity for any other structural mode unless explicitly detailed to do so.

9.6.6.4 Columns confined for increased ductility shall be detailed to achieve a ductility ratio, μ, of at least 6 for all LOP. Columns shall be designed so that the actual ductility demands for the design blast loads are:

LOP I and II, $\mu \leq 6$
LOP III, $\mu \leq 3$
LOP IV, $\mu \leq 1$

9.6.7 FRP Solid Sections. The following maximum strain criteria shall be utilized unless another failure criteria can be shown to be appropriate by analysis or testing:

$$\eta \varepsilon_{11}^{CU} \leq \varepsilon_{11} \leq \eta \varepsilon_{11}^{TU}$$
$$\eta \varepsilon_{22}^{CU} \leq \varepsilon_{22} \leq \eta \varepsilon_{22}^{TU}$$
$$|\varepsilon_{12}| \leq \eta \varepsilon_{12}^{SU}$$

where the superscripts cu, tu, and su of the strain components represent compressive ultimate, tensile ultimate, and shear ultimate strains, respectively.

9.6.7.1 For each LOP, the nondimensional factor, η, shall be:

LOP I and II, $\eta \leq 1.0$
LOP III, $\eta \leq 0.8$
LOP IV, $\eta \leq 0.5$

9.7 OTHER MATERIALS

Structural systems designed using materials identified below shall be designed to meet or exceed the provisions in the provided references.

9.7.1 Aluminum. The Aluminum Association's *Aluminum Design Manual* (AAI 2005).

The Aluminum Association's *Aluminum Standards and Data* (AAI 2006).

9.7.2 Wood. The American Forest & Paper Association's National Design Specification for Wood Construction, ANSI/AF&PA NDS-2005 (AF&PA 2005).

9.7.3 Cold-Formed Steel Framing. It is permitted to use cold-formed steel panels as part of the exterior wall system. Steel panels shall conform to ASTM standards as specified in the appropriate AISI references. Individual panels shall be designed to resist local buckling and web crippling. Response limits for cold-formed panels with and without membrane action are given in Table 3-2. An average strength factor and dynamic increase factor given in Section 3.5 shall be applied to the specified minimum yield stress.

Other framing systems such as wall studs and floor joists are not precluded by this standard if designed per AISI standards identified in this chapter (at a minimum) and capable of achieving the design intent for the LOP and response limits as given in Tables 3-1 and 3-2.

9.8 CONSENSUS STANDARDS AND OTHER REFERENCED DOCUMENTS

The following references are consensus standards and are to be considered part of these provisions to the extent referred to in this chapter:

ACI
American Concrete Institute
38800 Country Club Drive
Farmington Hills, MI 48331
Building Code Requirements for Structural Concrete, ACI 318, 2008.Building Code Requirements and Specifications for Masonry Structures, ACI 530/530.1-8; ASCE 5-08/6-08; TMS 402/602, 2008.
Guide for the Design and Construction of Externally Bonded FRP Systems for Strengthening Concrete Structures, ACI 440.2R-02, 2002.

AISC
American Institute of Steel Construction
One East Wacker Drive, Suite 700
Chicago, IL 60601-1802
Specification for Structural Steel Buildings, ANSI/AISC 360-05, 2005.
Seismic Provisions for Structural Steel Buildings, ANSI/AISC 341-05, 2005.

AISI
American Iron and Steel Institute
1140 Connecticut Ave., NW, Suite 705
Washington, DC, 2004 20036
North American Standard for Cold-Formed Steel Framing—Truss Design, AISI S214-07 with S2-08, 2007 ed. with Supplement 2, 2008.
North American Specification for the Design of Cold-Formed Steel Structural Members, AISI S100-07, 2007.
North American Standard for Cold-Formed Steel Framing—General Provisions, AISI S200-07, 2007.
North American Standard for Cold-Formed Steel Framing—Floor and Roof System Design, AISI S210-07, 2007.
North American Standard for Cold-Formed Steel Framing—Wall Stud Design, AISI S211-07, 2007.

North American Standard for Cold-Formed Steel Framing—
Header Design, AISI S212-07, 2007.

North American Standard for Cold-Formed Steel Framing—
Lateral Design, AISI S213-07, 2007.

Aluminum Association Inc. (AAI). (2006). Aluminum
Standards and Data. AAI, Arlington, Va.

AAI. (2005). Aluminum Design Manual. AAI, Arlington, Va.

American Forest & Paper Association (AF&PA). (2005).
National Design Specification for Wood Construction,
ANSI/AF&PA NDS-2005. AF&PA, Washington, D.C.

ASTM International (ASTM). (2006). Standard Specification
for Deformed and Plain Carbon Steel Bars for Concrete
Reinforcement, ASTM A615/A615M. ASTM, West
Conshohocken, Pa.

ASTM. (2006). Standard Specification for Low-Alloy Steel
Deformed and Plain Bars for Concrete Reinforcement,
A706/A706M-06a. ASTM, West Conshohocken, Pa.

Steel Deck Institute
P.O. Box 25
Fox River Grove, IL 60021
SDI Design Manual for Composite Decks, Form Decks and
Roof Decks, Publication No. 31, 2007.

U.S. Department of Defense
Structures to Resist the Effects of Accidental Explosions,
UFC 3-340-02, 2008.

Wire Reinforcement Institute
942 Main Street, Suite 300
Hartford, CT 06103
Manual of Standard Practice—Structural Welded
Reinforcement, WWR-500, 2006.
Structural Detailing Manual, WWR-600, 2006.

Chapter 10
PERFORMANCE QUALIFICATION

10.1 SCOPE

This chapter of the Standard provides procedures and standards for the peer review of the design of protected structures and the verification of performance of selected security-related components and devices. Reference is made in the Commentary to consensus standards and documents prepared by others. The minimum performance levels, standoff distances, and explosive weights presented in these standards and documents vary and the responsible design professional shall adjust these variables as needed.

10.2 PEER REVIEW

A third-party review of the analysis, design, and detailing of a protected building, and/or component or equipment qualification data, shall be performed at the discretion of the Authority Having Jurisdiction.

10.3 SITE PERIMETER COMPONENTS

It shall be permissible to qualify the performance of site perimeter components such as barriers, anti-ram devices, sculptures, and street landscaping by full-scale blast testing, analysis, and design, or a combination of analysis, design, and testing.

10.3.1 Performance Qualification by Full-Scale Testing. It shall be permissible to qualify the site perimeter components by full-scale testing for resistance to vehicular impact. The following information shall be reported:

1. Date(s) and times of testing and date of the report
2. Description of the component to be tested (hereafter termed the *test article*), including drawings of all structural, mechanical, and electrical components that influence impact resistance; test article embedment/foundation/ anchorage data; and properties of the construction and geomedia underlying the test article and test vehicle
3. Description of the impacting vehicle, including curb-side weight
4. Number of test articles
5. Ambient temperature within 5 min. of testing
6. Pre- and posttest photographs and high-speed test video
7. Impact conditions in terms of the measured speed, computed equivalent speed, and angle of attack of the test vehicle. (The measured speed is that of the test vehicle at impact. The computed equivalent speed shall be based on the energy absorbed by the test article and the mass of the test vehicle.)
8. Response data in sufficient detail to completely characterize the response of the test article and the impacting vehicle to enable calculation of the energy absorbed by the soil and foundation, the impacting vehicle, and the test article

9. Damage to the test article and test vehicle, test article transient and permanent deflection, postevent functionality of the test article, and test vehicle penetration.

10.3.2 Performance Qualification by Analysis and Design. It shall be permissible to qualify the performance of site perimeter components by analysis and design using computer codes (finite element analysis or hydrocode) and elements that have been validated using the results of vehicular impact tests if those parts of the site-perimeter components and the test vehicle that are not amenable to structural analysis have been characterized by appropriate full-scale testing.

Performance qualification of site-perimeter components by analysis and design shall be peer reviewed per Section 10.2 if required by the Authority Having Jurisdiction.

The mathematical model of the site-perimeter structure shall include the component foundation, the test vehicle roadway, and the geomedia beneath the component, its foundation, and the test vehicle. Nonlinear models shall be used for all elements not responding in the linear range under the design impact.

The following information shall be reported:

1. Date of the report
2. Complete description of the test article or shielding component, including drawings of all structural, mechanical, and electrical components that influence impact resistance; test article embedment/foundation/anchorage data; and properties of the construction and geomedia underlying the test article
3. Complete description of the impacting vehicle, including curb-side weight
4. The mathematical model of the test article and test vehicle
5. Impact conditions, including impact speed and angle of attack on the test article
6. Response data in sufficient detail to completely characterize the response of the test article and the impacting vehicle to enable calculation of the energy absorbed by the soil and foundation, the impacting vehicle, and the test article
7. Estimated damage to the test article, test article transient and permanent deflection, and test vehicle penetration.

10.4 BUILDING STRUCTURAL COMPONENTS

10.4.1 Performance Qualification by Full-Scale Testing. It shall be permissible to qualify the performance of components of building structures for resistance to blast loading by full-scale testing. The following information shall be reported:

1. Date(s) and times of testing and date of the report
2. Description of the tested components, test fixture, and component-to-fixture anchorages, including drawings,

construction specifications (e.g., welding procedure specifications) and material properties

3. Complete description of the charge, including TNT equivalent, geometry, casing (if any), standoff distance, distance above the ground, test fixture elevation (altitude), ambient temperature, and orientation with respect to the test article
4. Number of test articles
5. Pre- and posttest photographs and high-speed test video
6. Response data in sufficient detail to completely characterize the responses of the components and the measured pressure and impulse distribution over the face of the components
7. Damage to the components, component transient and permanent deflections, postevent functionality of components, and fragment penetration.

10.4.2 Performance Qualification by Analysis and Design. It shall be permissible to qualify the performance of structural components by analysis and design using computer codes (finite element analysis or hydrocode) and elements that have been validated using the results of blast and fragmentation tests if those parts of the structural components that are not amenable to analysis have been characterized by appropriate full-scale testing. Nonlinear models shall be used.

Performance qualification of a structural component by analysis and design shall be peer reviewed per Section 10.2 if required by the Authority Having Jurisdiction.

The following information shall be reported:

1. Date of the report
2. Results of prior validation studies to support the use of the finite element code and its constitutive models
3. Complete description of the material models, the test component, and the assumed boundary conditions
4. Complete description of the charge, standoff distance, and elevation
5. The mathematical model of the component
6. Blast loading pressure histories across the face of the component
7. Response data in sufficient detail to completely characterize the response of the component
8. Estimated damage to the component, component transient and permanent deflection, postevent functionality of the component, and fragment penetration.

10.5 SHIELDING STRUCTURES

10.5.1 Performance Qualification by Full-Scale Testing. It shall be permissible to qualify the performance of shielding structures (e.g., blast walls) for resistance to blast loading by full-scale testing. The following information shall be reported:

1. Date(s) and times of testing and date of the report
2. Description of the test article or component, including drawings of all structural, mechanical, and electrical components that influence blast resistance; test article embedment/foundation /anchorage data; and properties of the geomedia underlying the test article
3. Complete description of the charge, including TNT equivalent, geometry, casing (if any), standoff distance, distance above the ground, test fixture elevation (altitude), ambient temperature, and orientation with respect to the test article
4. Number of test articles
5. Ambient temperature within 5 min. of testing
6. Pre- and posttest photographs and high-speed test video

7. Response data in sufficient detail to completely characterize the response of the test article and the measured pressure and impulse distribution on both faces of the test article to determine net reductions in blast loading at the test article
8. Damage to the test article, test article transient and permanent deflection, postevent functionality of the test article, and fragment penetration.

10.5.2 Performance Qualification by Analysis and Design. It shall be permissible to qualify the performance of shielding structures such as blast walls by analysis and design using computer codes (finite element analysis or hydrocode) and elements that have been validated using results of blast and fragmentation tests if those parts of the shielding structure that are not amenable to structural analysis have been characterized by appropriate full-scale testing.

Performance qualification of a shielding structure by analysis and design shall be peer reviewed per Section 10.2 if required by the Authority Having Jurisdiction.

The mathematical model shall include the structure, its foundation, and the subsurface materials beneath the structure. Nonlinear models shall be used for all elements not responding in the linear range under the design blast loading.

The following information shall be reported:

1. Date of the report
2. Description of the test article, including drawings of all structural, mechanical, and electrical components that influence blast resistance; test article embedment/foundation/ anchorage data; and properties of the construction and geomedia underlying the test article
3. Complete description of the charge, standoff distance, and elevation
4. The mathematical model of the test article
5. Blast loading pressure histories across the face of the shielding component
6. Response data in sufficient detail to completely characterize the response of the test article to enable calculation of the energy absorbed by the soil, foundation, and test article
7. Estimated damage to the test article, test article transient and permanent deflection, postevent functionality of the test article, and fragment penetration.

10.6 BUILDING FAÇADE COMPONENTS

10.6.1 Glazing and Glazing Systems
10.6.1.1 Performance Qualification by Full-Scale Testing. It shall be permissible to qualify the performance of glazing, glazing systems (including curtain walls), and glazing retrofit systems by full-scale testing, including those fabricated from glass, glass-clad plastics, laminated glass, glass/plastic glazing materials, and film-backed glass. The requirements below shall be followed.

1. A minimum of three test specimens representative of a glazing, glazing system (excluding curtain walls), or glazing retrofit system shall be tested for a given combination of airblast loading that is defined using a reflected pressure and a positive phase impulse. A minimum of one curtain wall specimen shall be tested.
2. Testing shall be conducted in either an open-air arena or a shock tube. The test specimen shall be installed in the test frame perpendicular to a line from the point of detonation to the center of the test frame.

3. If an explosive charge is used for testing, the charge shall be hemispherical and detonated either at ground level or elevated by placing the charge on a table top that is between 24 in. (0.61 m) and 48 in. (1.2 m) above the ground. Other explosive charge configurations may be used but the effects of the alternate charge configuration must be documented. If there is a possibility of crater ejecta (fragments) altering the results of a test, the explosive charge shall be placed on a blast mat.
4. The test frame shall be part of the test facility. The face of the test frame with the test specimen installed shall be a plane surface. No openings shall exist in the front face of the test frame so as to prevent airblast leakage behind the test specimen.
5. Calibrated transducers and high-speed digital cameras shall be used to record the demands on and the response of the test specimen.

The following information shall be reported:

1. Date(s) and times of testing and date of the report
2. Description of the test specimen, including manufacturer, source of supply, dimensions, model number, materials, detailed drawings of the specimen (e.g., section profiles, framing layout, test panel arrangement, locking and hinge arrangements, sealants) and locations of all transducers
3. Complete description of the test fixture (arena test and shock tube)
4. Number of specimens tested
5. Ambient temperature and the temperature of the glazing, both within 5 min. of testing
6. Pre- and posttest photographs and high-speed test video
7. Airblast pressure history, peak positive pressure, positive phase duration and positive phase impulse measured from each reflected airblast pressure transducer, peak negative pressure, negative phase duration and negative phase impulse measured from each reflected airblast pressure transducer
8. Complete description of the charge, standoff distance, and elevation
9. Damage to the test specimen, test specimen transient and permanent deflection, postevent functionality of the test specimen, and fragment penetration.

It shall be permissible to use the procedures and requirements for testing glazing components and glazing systems (including curtain walls) presented in ASTM F1642-04 (ASTM 2010) for determining and reporting the performance of a glazing, glazing system or glazing system retrofit subjected to airblast loading.

10.6.1.2 Performance Qualification by Analysis and Design. It shall be permissible to qualify the performance of glazing and glazing systems by analysis and design using computer codes (finite element analysis or hydrocode) and elements that have been validated using results of blast tests if those parts of the glazing system that are not amenable to structural analysis have been characterized by appropriate full-scale testing.

Performance qualification of glazing and glazing systems by analysis and design shall be peer reviewed per Section 10.2 if required by the Authority Having Jurisdiction.

The finite element model shall include all components of the glazing system and appropriately account for boundary conditions and contact between the glazing panels, support framing, mullions, etc. Nonlinear models shall be used for all elements not responding in the linear range under the blast loading.

The following information shall be reported:

1. Date of the report
2. Description of the test article or component, including drawings of all structural and mechanical components that influence blast resistance
3. Complete description of the charge, standoff distance, and elevation
4. The mathematical model of the test article
5. Blast loading pressure histories across the face of the test article
6. Response data in sufficient detail to completely characterize the response of the test article.
7. Estimated damage to the test article, test article transient and permanent deflection, and postevent functionality of the test article.

10.6.2 Doors. Blast-resistant doors are designed to protect against the effects of internal or external airblasts. The performance objectives for a blast door can include (1) no pressure leakage during or after testing, (2) no structural damage during and after testing, and (3) operability after testing.

10.6.2.1 Performance Qualification by Full-Scale Testing. It shall be permissible to qualify the performance of doors by full-scale blast testing provided that the following requirements are satisfied:

1. A minimum of three test specimens shall be tested for a given combination of airblast loading that is defined using a reflected pressure and a positive phase impulse.
2. Testing shall be conducted in either an open-air arena or a shock tube. The test specimen shall be installed in the test frame perpendicular to a line from the point of detonation to the center of the test frame.
3. If an explosive charge is used for testing, the charge shall be hemispherical and detonated either at ground level or elevated by placing the charge on a table top that is between 24 in. (0.61 m) and 48 in. (1.2 m) above the ground. Other explosive charge configurations may be used but the effects of the alternate charge configuration must be documented. If there is a possibility of crater ejecta (fragments) altering the results of a test, the explosive charge shall be placed on a blast mat.
4. The test frame shall be part of the closed test facility. The face of the test frame with the test specimen installed shall be a plane surface. No openings shall exist in the front face of the test frame so as to prevent airblast leakage behind the test specimen.
5. Calibrated transducers and high-speed digital cameras shall be used to record the demands on and the response of the test specimen.

The following information shall be reported:

1. Date(s) and times of testing and date of the report
2. Description of the test specimen, including manufacturer, source of supply, dimensions, model number, materials, detailed drawings of the specimen (e.g., section profiles, door dimensions, framing layout, test panel arrangement, locking and hinge arrangements, sealants) and locations of all transducers
3. Complete description of the test fixture (arena test and shock tube)
4. Number of specimens tested
5. Ambient temperature within 5 min. of testing
6. Pre- and posttest photographs and high-speed test video
7. Airblast pressure history, peak positive pressure, positive phase duration and positive phase impulse measured from

each reflected airblast pressure transducer, peak negative pressure, negative phase duration and negative phase impulse measured from each reflected airblast pressure transducer

8. Complete description of the charge, standoff distance, and elevation
9. Damage to the test specimen, test specimen transient and permanent deflection, post-event functionality of the test specimen, and fragment penetration.

It shall be permissible to use the procedures and requirements in ASTM F2247-11 for determining and reporting the performance of metal doors subjected to airblast loading.

10.6.2.2 Performance Qualification by Analysis and Design. It shall be permissible to qualify the performance of blast doors by analysis and design using computer codes (finite element analysis or hydrocode) and elements that have been validated using results of blast tests if those parts of the blast door that are not amenable to structural analysis have been characterized by appropriate full-scale testing.

Performance qualification of blast doors by analysis and design shall be peer reviewed per Section 10.2 if required by the Authority Having Jurisdiction.

The mathematical model shall include the all components of the blast door and appropriately account for boundary conditions and contact between the door and support framing. Nonlinear models shall be used for all elements not responding in the linear range under the blast loading.

The following information shall be reported:

1. Date of the report
2. Description of the test article or component, including drawings of all structural and mechanical components that influence blast resistance
3. Complete description of the charge, standoff distance, and elevation
4. The mathematical model of the test article
5. Blast loading pressure histories across the face of the test article
6. Response data in sufficient detail to completely characterize the response of the test article
7. Estimated damage to the test article, test article transient and permanent deflection, and postevent functionality of the test article.

10.7 BUILDING NONSTRUCTURAL COMPONENTS

10.7.1 Performance Qualification by Full-Scale Testing. It shall be permissible to qualify the performance of nonstructural components or equipment for resistance to blast loading by full-scale testing. The following information shall be reported:

1. Date(s) and times of testing and date of the report
2. Description of the test article or component, including drawings of all structural, mechanical, and electrical components that influence blast resistance
3. Complete description of the charge, including TNT equivalent, geometry, casing (if any), standoff distance, distance above the ground, test fixture elevation (altitude), ambient temperature, and orientation with respect to the test article
4. Number of test articles
5. Ambient temperature within 5 min. of testing
6. Pre- and posttest photographs and high-speed test video
7. Response data in sufficient detail to completely characterize the response of the test article and the pressure and impulse distribution over the face of the test article

8. Damage to the test article, test article transient and permanent deflection, postevent functionality of the test article, and fragment penetration.

It shall be permissible to qualify the performance of nonstructural components or equipment for resistance to ground or in-structure loading by full-scale shock testing. The following information shall be reported:

1. Date(s) and times of testing and date of the report
2. Description of the test article or component, including drawings of all structural, mechanical, and electrical components that influence blast resistance
3. Complete description of the applied shock load, including method of application, impact location, orientation of the test article, and maximum acceleration
4. Number of test articles
5. Pre- and posttest photographs and high-speed test video
6. Response data in sufficient detail to completely characterize the response of the test article
7. Damage to the test article, test article transient and permanent deflection, test article maximum acceleration, and postevent functionality of the test article.

10.7.2 Performance Qualification by Analysis and Design. It shall be permissible to qualify the performance of nonstructural components and equipment by analysis and design using computer codes (finite element analysis or hydrocode) and elements that have been validated using results of blast tests if those parts of the nonstructural components or equipment that are not amenable to structural analysis have been characterized by appropriate full-scale testing.

The mathematical model shall include the mass and flexibility of all components, including their relative position and connectivity. Nonlinear models shall be used for all elements not responding in the linear range under the design blast loading.

Performance qualification of a nonstructural component by analysis and design shall be peer reviewed per Section 10.2 if required by the Authority Having Jurisdiction.

The following information shall be reported:

1. Date of the report
2. Description of the test article or component, including drawings of all structural, mechanical, and electrical components that influence blast resistance
3. Complete description of the charge, standoff distance, and elevation
4. The mathematical model of the test article
5. Blast loading pressure histories across the face of the component
6. Response data in sufficient detail to completely characterize the response of the test article
7. Estimated damage to the test article, test article transient and permanent deformations, and postevent functionality of the component.

10.8 CONSENSUS STANDARDS AND OTHER REFERENCED DOCUMENTS

ASTM International
100 Barr Harbor Drive
P.O. Box C700
West Conshohocken, PA 19428-2959
Standard Test Method for Metal Doors Used in Blast Resistant Application, ASTM F2247-11, 2011.
Standard Test Method for Glazing and Glazing Systems Subject to Airblast Loadings, ASTM F1642-04, 2010.

COMMENTARY

This Commentary is not a part of the mandatory section of the ASCE Standard for Blast Protection of Buildings (the Standard). It is included for information purposes.

This Commentary consists of explanatory and supplementary material designed to assist users of the Standard in applying its requirements. It is intended to create a better understanding of these requirements through brief explanations of the reasoning employed in arriving at them.

The sections of the Commentary are numbered to correspond to the sections of the Standard to which they refer. Since it is not necessary to have supplementary material for every section in the Standard, there are gaps in the numbering of the Commentary.

Chapter C1
GENERAL

C1.1 SCOPE

Considerations for various structural and nonstructural measures to mitigate blast effects are included in this Standard. However, this Standard does not prescribe which buildings should be subject to its provisions, nor the size of explosive or specific performance criteria for a given situation. These decisions are intended to be the responsibility of the building owner, manager, or tenant, informed by consultation with a qualified user as defined in Section 1.4.

Certain U.S. Department of Defense (DoD) facilities are required to comply with the uniform explosives safety regulations specified in DoD 6055.09-STD (DoD 2009), and other facilities where ammunition or explosives are routinely present should conform to these or comparable requirements. Design of protective structures in such cases should be in accordance with UFC 3-340-02 (DoD 2008), rather than this Standard.

The scope of this Standard is generally limited to the evaluation of blast effects on structural and nonstructural elements and systems and does not include the evaluation of the subsequent behavior of a damaged structure, such as the potential for progressive collapse.

C1.4 QUALIFICATIONS

The statutes and administrative laws governing the practice of engineering in most United States jurisdictions include consultation, investigation, and evaluation—not just planning and design—within the scope of regulated activity for which licensure is legally mandated. Individual building owners, such as federal agencies, may establish their own minimum qualifications for individuals who perform blast effects analysis of their facilities.

REFERENCES

U.S. Department of Defense (DoD). (2009). DoD Ammunition and Explosives Safety Standards, DoD 6055.09-STD, Incorporating Change 2, August 21, 2009, <www.ddesb.pentagon.mil/2009-08-21%20-%20 (Change%202%20to%2029%20Feb%2009%20Version)%20DoD%20 6055.09-STD,%20DoD%20AE%20Safety%20Standards.pdf> [May 12, 2011].

DoD. (2008). Structures to Resist the Effects of Accidental Explosions, UFC 3-340-02, <www.wbdg.org/ccb/DOD/UFC/ufc_3_340_02.pdf> [May 12, 2011].

Chapter C2
DESIGN CONSIDERATIONS

C2.1 SCOPE

This chapter is intended to be a guideline that private-sector owners and their consultants can use to establish an appropriate scope of work for projects in this area. This chapter covers general conceptual issues, while subsequent chapters address specific design considerations. This chapter does not stipulate a particular size of explosive, but presents guidance on how to go about selecting one for a given situation. Also, this chapter does not specify a particular risk assessment methodology; rather, it provides basic principles and a framework for such a process.

C2.2 RISK ASSESSMENT

There are two approaches for establishing the appropriate structural design criteria for mitigating blast effects on a particular building. In some cases, requirements such as the size and location of the explosive and the acceptable response of building elements are prescribed by an Authority Having Jurisdiction. However, it is more prevalent for these parameters to be established on the basis of an assessment of the risk associated with a nearby explosion. This risk is traditionally defined as the combination (product) of three components: consequence, threat, and vulnerability.

There are many valid risk assessment methodologies in widespread use for various types of buildings and infrastructure. Examples include API RP 752 (API 2003) for accidental threats, and FEMA 452 (FEMA 2005) and UFC 4-020-01 (DoD 2007b) for malicious threats.

Risk changes over time. Consequently, a revised risk assessment is recommended at periodic intervals—for example, every 5 years—as well as under any of the following circumstances:

- A change of the mission and/or assets housed in the building
- A change in the building's threat environment
- Significant physical modification of the building itself
- Construction of a neighboring facility.

C2.2.1 Consequence Analysis. Relevant background information about the facility will include the number and type of tenants; the number of employees and visitors; the mission of each tenant; the area, by location and size, occupied by each tenant; and a physical description of the construction elements of the facility.

Except for certain critical facilities, the objective of design is to protect the occupants and contents of the building, not the structure itself. Consequence analysis is therefore directed at the assets—people, property, and information—that are housed in the building and the impact of their loss or compromise on the mission of the owner or users. These assets must be identified and then assigned a weight on the basis of such factors as criticality, replacement time, replacement cost, and quantity.

Building codes typically differentiate between "ordinary" occupancies and those that warrant greater protection from environmental effects. Examples of such structures include:

- Facilities where a large number of people congregate in one area
- Schools and daycare facilities where a large number of children are present
- Colleges and other education facilities where a large number of adult students are present
- Hospitals and other health care facilities
- Power generating stations and other public utility facilities
- Facilities that manufacture, process, handle, store, use, or dispose of significant quantities of such substances as hazardous fuels, hazardous chemicals, hazardous waste, or explosives
- Fire, rescue, ambulance, and police stations and emergency vehicle garages
- Designated earthquake, hurricane, or other emergency shelters
- Designated emergency preparedness, communication, and operation centers and other facilities required for emergency response
- Aviation control towers, air traffic control centers, and emergency aircraft hangars
- Water storage facilities and pump structures required to maintain water pressure for fire suppression
- Buildings and other structures having critical corporate, government, or national defense functions.

The same is true when designing for an explosion. Buildings that are occupied by a relatively large number of people or have occupants concentrated within a relatively small area have higher consequences of damage and failure than less populated or less densely populated structures such as industrial, maintenance, and storage facilities.

C2.2.2 Threat Analysis. Explosive threats to be used for blast-resistant design are typically classified as either accidental or intentional. Accidental threats are usually associated with explosive materials that are stored and handled at or near the facility. Intentional threats are military or improvised devices, which may be located inside a moving or parked vehicle, a suitcase or briefcase, or other place of concealment and delivery.

An increasingly common strategy in blast-resistant design is to specify at least two different design basis threats: a relatively small explosive against which the building must provide a relatively high level of protection, and a larger explosive against which it is acceptable for the building to provide a lower level of protection. This is analogous to performance-based seismic design, which typically aims for immediate occupancy in a mild earthquake and life safety in an extreme earthquake. This approach provides the building owner with a better understanding and expectation of performance in response to an inherently low-probability, high-consequence event.

C2.2.2.2 Malicious Threats. Security threats are acts or conditions that may result in loss of life; damage, loss, or destruction of property; or disruption of mission. Physical security personnel and design teams must understand the threat to the facilities that they are tasked to protect in order to develop effective security programs or designs, security systems, and/or structural upgrades. Historical patterns and trends in threat activity indicate general categories of threats and the common tactics that are used against facilities. Threat tactics and their associated tools, weapons, and explosives are the basis for the threat to facilities.

Criminal activities in the area may increase the likelihood of some threats. An identification of possible threats in the area should be obtained from federal, state, or local law enforcement officials. Within the United States and its territories, the Federal Bureau of Investigation (FBI) has primary responsibility for both foreign and domestic terrorists. The FBI and state and local law enforcement agencies are good sources from which to obtain criminal threat information. The possible threats, along with their preferred method(s) of attack, should be available from these agencies.

The threat must be described in specific terms to help determine the facilities' vulnerabilities or to establish protective measures. This description should include the tactics that aggressors will use to compromise the facility, as well as any weapons, tools, or explosives that are likely to be used in an attempt. For example, the threat might be described as a moving vehicle bomb consisting of a 4,000-lb (1,800-kg) vehicle containing a 500-lb (225-kg) explosive traveling at 30 mph (13 m/s).

The likelihood of the threat is analyzed for each applicable threat category by considering, in turn, five factors: the existence of an aggressor; the capabilities of the aggressor; the history of attacks by similar aggressors against similar assets; the intentions of the aggressor; and any intelligence indicating that the aggressor is actively targeting this facility (FEMA 2005). Threat likelihood is often correlated with threat magnitude based on the principles of risk acceptance; if the likelihood of an attack is low, the building can be designed for a less severe threat under the assumption that the aggressor will expend less effort and fewer resources on targets that are less attractive (DoD 2007b). However, unlike natural hazards, it is not feasible to develop a truly probabilistic approach to establishing design loads for intentional threats because the initiator is not a physical process, but, rather, an intelligent adversary.

C2.2.3 Vulnerability Analysis. Vulnerabilities are weaknesses in the facility's protective systems. They are identified by considering the tactics associated with the identified threat and the levels of protection directed against those tactics. Some vulnerabilities can be identified by considering the general design strategies for each adversarial tactic (DoD 2007b). The general design strategies identify the basic approach to protecting assets against specific tactics. For example, the general design strategy for a moving vehicle bomb is to keep the vehicle as far from the facility as possible and to harden the facility to resist the explosive at that distance. Examples of potential vulnerabilities include limited standoff distances, inadequate barriers, and building construction that cannot resist explosive effects at the achievable standoff distance.

C2.2.4 Risk Analysis. The prioritization of protection should be based on both the results of the risk assessment (Section 2.2) and the building owner's risk acceptance decisions (Section 2.4).

C2.3 RISK REDUCTION

Structural hardening to reduce vulnerability is often not the most cost-effective approach for protecting a building from blast effects. A variety of physical and operational measures should also be considered (API 2003; FEMA 2003a).

C2.3.1 Consequence Reduction. Asset redundancy reduces the impact should one of them be lost or compromised. Asset dispersal reduces the impact of an explosion near one of them.

Site selection is a key early step in the design process for a new facility. The effects of an accidental explosion can be mitigated simply by locating the building as far as possible from all possible sources of such an event. Similarly, consideration should be given when siting a new structure to the possibility of a bomb attack against a nearby target. Choosing an isolated location will minimize the potential for collateral damage.

It is possible to reduce the impact of an explosion by having policies and procedures in place for maintaining operations in the event of localized damage or shifting functions elsewhere if the building as a whole is affected. The facility should be equipped with a mass notification system to ensure that occupants can be advised of threats and given instructions about where to go and what to do (DoD 2007a).

C2.3.1.1 Nonstructural Components and Systems. Essential systems typically include emergency power or lighting and all utilities serving safe havens. It is unlikely that standard components will be able to withstand direct blast pressures. Careful positioning within the structure can reduce the loads, generally resulting in the most cost-effective solution. Special care must be taken to ensure that services required for continued operation of essential components and systems, such as power and water, are also designed to remain in operation, or that redundant systems are provided.

Design of nonessential components and systems is often addressed by specifying equivalent seismic loads. Great care must be taken with this approach, as the seismic loads specified often exempt many of the nonstructural items. It is preferable to specify the actual loads to be resisted, often as a multiple of gravity (DoD 2007a), or to specify a particular seismic design category that will result in loads of the desired magnitude. This is often not the code-mandated seismic design category for the project location. There are industry-specific guidelines for seismic design that can also provide a convenient method for specifying the design loads (SMACNA 1998).

The performance of essential nonstructural components and systems must be qualified by either full-scale testing or the use of data acquired from other sources. If the operational portions of the equipment or components are not directly exposed to blast loads, either because of their location within the building or by encasement in properly designed enclosures, then they need be qualified only for blast-induced ground or in-structure shock or the specified lateral and/or vertical loads. Nonessential components and systems may be qualified by either full-scale testing, experimental data, or analysis to ensure position retention. Qualification for a specific installation should include evaluation of anchorage and the support structure.

C2.3.2 Threat Reduction
C2.3.2.1 Accidental Threats. Accidental explosions typically occur when materials stored in or near a facility are inadvertently detonated or deflagrated. Items in these categories could include paints and paint thinners, gasoline, propane and similar gases, liquid natural gas (LNG), certain kinds of dust, fireworks, explosive charges and igniters, and cleaning solvents. Whenever possible, these items should be secured and stored in a separate

building away from the main facility. If they have to be stored inside of the facility, the room storing these items should be designed to withstand the associated explosive force or vent it to the exterior to mitigate the risk of harm to the facility (NFPA 2007). Tenants should also develop and enforce appropriate policies and procedures for safe handling of hazardous materials (NFPA 2008).

C2.3.2.2 Malicious Threats. Malicious explosions are, by definition, the result of deliberate actions by intelligent human aggressors. In addition to standoff, discussed below, effective means of reducing the threat of such an attack include physical and operational security.

One approach to physical security involves creating a layered defensive system to deter hostile acts and prevent or delay access to a building. Such a system can include as many as five tiers: guarded perimeter, standoff zone, building exterior, interior access control, and safe havens. The complete system must be three-dimensional in construction and address the space above, below, and around the building to be protected.

Physical security is most effectively achieved when the asset being protected is in the center of the defensive system, with the five tiers representing concentric layers of detection, delay, and response. Detection elements are capable of sensing the presence of unauthorized individuals or explosives and alerting security personnel accordingly. Delay elements are physical boundaries such as fences, gates, walls, doors, roofs, ceilings, etc. that take time for an aggressor to penetrate. Response elements are typically guard forces charged with intercepting aggressors after they have been detected, but before an attack can be carried out.

Detection elements should be installed adjacent to barriers to increase the probability of sensing an intruder. Detection elements should be mutually supportive and within the coverage of cameras, protective lighting, or other alarm assessment devices to facilitate response.

Visible security measures may serve as a deterrent to an attack since they make obvious the difficulty of carrying one out successfully. Vehicle searches at the site perimeter and personal searches at the building entrance limit the size of explosive that can be brought near and inside the facility, respectively. The effectiveness of such searches will be enhanced if they include gamma-ray inspection of cargo, x-ray screening of bags, magnetometer screening of individuals, and/or use of explosive trace detectors. Finally, access control for vehicles and/or persons will ensure that only those with appropriate authorization can approach and/or enter the building.

The following exterior security measures should be considered and applied as appropriate: perimeter barriers preventing unauthorized entry to the site; site layout to achieve maximum standoff in every direction, provide difficult approach zones, and limit areas of potential concealment within a particular distance around the building; separation between the building and other buildings, parking areas, public areas, and public transportation; guard stations; and control of vehicle access, such as eliminating lines of approach perpendicular to the building and restricting parking to authorized vehicles only.

Site design should include, to the greatest extent possible, the principles of Crime Prevention Through Environmental Design (CPTED). The four main concepts of CPTED are natural access controls such as properly located entrances, exits, and fences; target hardening such as access controls, surveillance cameras, and barriers; territorial reinforcement such as clearly demarcated boundaries; and natural surveillance such as eliminating visual obstructions.

Other specific elements in the guarded perimeter and standoff zones include clear zones, security lighting, intrusion detection devices, alarm systems and sensors, and alert and notification systems. These mutually supportive elements should be integrated with elements described elsewhere in this Standard, as well as the overall development of the building site, including landscaping, parking, roads, and other features. Standoff should be coupled with appropriate building details and site features wherever possible to provide the desired level of protection.

No guidance is provided herein to mitigate attacks with artillery-type munitions such as mortars or rockets. Direct-fire weapons that require a line of sight can be thwarted with predetonation screens and walls.

Security measures for the building exterior zone should account for all points of entry and exit for employees, visitors, and utility services to a building. Public access should be limited to the minimum number of entrances consistent with functional and life safety requirements. Service and utility entrance and exit points—such as air intakes, mechanical ducts, roof hatches, and water supplies—should be adequately safeguarded. Communications systems within the site should not be unduly vulnerable to accidental or intentional disruption.

For interior access control, the following security measures should be considered and applied as appropriate: building layout and compartmentalization; interior construction, including circulation routes and locations of elevators, stairwells, and entry control points; creation of well-defined and secured zones for controlling the flow of employees, contractors, visitors, and service personnel; securing of utility closets, mechanical rooms, and access doors to duct shafts and ceiling spaces; protection of nonstructural elements and systems that are necessary to fulfill the building's operational and functional requirements; and location of critical building systems away from areas considered to be potential targets, as well as all public access zones.

Operational security includes policies and procedures within the building or organization. Considerations include activities of guards, employees, visitors, and service personnel within the building. The following measures should be considered and applied as appropriate: coordinating central command center and satellite posts; maintaining building systems monitoring and control capabilities; implementing shipping and delivery procedures, especially for an underground parking facility or loading dock; and personnel/visitor screening and monitoring at the entrances.

C2.3.2.2.1 Standoff. It is desirable to keep the threat as far away from the building as possible. As described in Chapter 4, the effects of an explosion decrease rapidly with the distance between the source and the target. Increasing standoff is often the most cost-effective way to reduce potential vulnerability and associated risk, regardless of the assumed size of an explosive charge. Maximizing standoff also ensures that there is opportunity in the future to upgrade the building for increased threats or a higher level of protection or performance.

The standoff distance necessary to avoid hardening the building is a function of the type and weight of the explosive, the type of construction of the building, and the desired level of protection (DoD 2007b). Figures C2-1 and C2-2 indicate the minimum standoffs required by the DoD for inhabited buildings of conventional construction, i.e., without specific hardening for blast effects. It is important to recognize that these distances are calibrated to specific explosive sizes, the details of which constitute sensitive information that should not be made available to the general public (DoD 2007a).

Where standoff is limited, the presence of blast walls between the building and the assumed location of the explosion may be beneficial. However, blast walls typically have limited use in

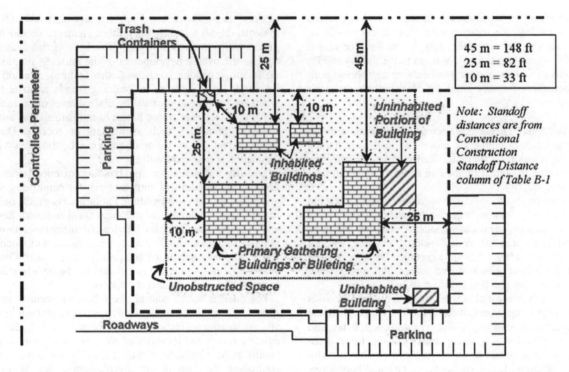

FIGURE C2-1. DoD MINIMUM STANDOFFS FOR BUILDINGS OF CONVENTIONAL CONSTRUCTION WITH A CONTROLLED PERIMETER (DoD 2007a).

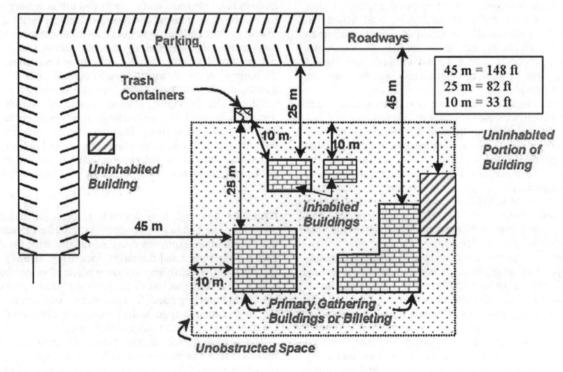

FIGURE C2-2. DoD MINIMUM STANDOFFS FOR BUILDINGS OF CONVENTIONAL CONSTRUCTION WITHOUT A CONTROLLED PERIMETER (DoD 2007a).

protective design because their effectiveness is highly dependent on the charge type and size and the distances between the explosive, wall, and building. Blast-induced shockwaves, although disrupted by the presence of a blast wall, can re-form behind the wall and may actually lead to greater localized demands than if the wall were not present. Test data suggest that blast walls with scaled heights of 0.8 ft/lb$^{1/3}$ (0.3 m/kg$^{1/3}$) to 2.0 ft/lb$^{1/3}$ (0.8 m/kg$^{1/3}$) will reduce the reflected pressure and impulse on building surfaces if they are located within a scaled distance of 3.0 ft/lb$^{1/3}$ (1.2 m/kg$^{1/3}$) from the explosive and a scaled distance between 1.0 ft/lb$^{1/3}$ (0.4 m/kg$^{1/3}$) and 20 ft/lb$^{1/3}$ (8.0 m/kg$^{1/3}$) from the building, although hardening is often more economical at scaled distances greater than 10 ft/lb$^{1/3}$ (4.0 m/kg$^{1/3}$). In all cases, values are referenced to the TNT-equivalent size of the explosive. Guidance for the adjustment of building loads when blast walls satisfying these constraints are present is provided in TM 5-853-3 (DoD 1994). Alternatively, computational fluid dynamics (CFD) tools can be used to characterize pressure distributions in the presence of blast walls.

Blast walls have been constructed from reinforced concrete, FRP composites, wood, and other materials. Some types of blast walls can mitigate the effects of fragmentation, even when the reduction of airblast effects is minimal. On the other hand, blast walls can also be the source of large secondary fragments that could be damaging to the structure that the wall is intended to protect. Consideration should be given to the construction of earth fill or other landforms, such as berms, behind the blast wall to capture these fragments, such as those caused by breach or spalling of reinforced concrete when the wall thickness is not sufficient to prevent these types of failure. Guidance for the design and construction of blast walls is provided in UFC 3-340-02 (DoD 2008).

C2.3.2.2.2 Vehicle Barriers. Perimeter barriers are required to enforce standoff for malicious explosions. The threat of a suicide attack using a moving vehicle requires the use of passive and/or active anti-ram structures, such as bollards, special planters, knee walls, gates, or pop-up devices. When only stationary vehicle bombs are being considered, it is usually sufficient to keep parking and roadways away from the building, especially if some kind of access control is in place. Hand-carried bombs can be hidden in trash containers, yard equipment, or even landscaping features. Therefore, the building should be surrounded by an unobstructed space that provides no opportunities for concealment of a small explosive device.

Most vehicle barriers rely on substantial foundations and ductile elements to absorb the kinetic energy of a moving vehicle. In many cases, the capacity of the barrier depends heavily on the construction of the foundation; therefore, the barrier and foundation must be considered as a single system. Surface-mounted barriers, such as blocks of concrete and boulders, rely primarily on their weight and friction, but may require a 1-in. (25-mm) embedment in the pavement or sidewalk to provide adequate resistance to sliding to limit vehicle penetration.

A maximum clear distance between an active anti-ram structure and an adjacent passive vehicle barrier is specified in order to prevent small vehicles from penetrating the barrier system. A minimum height for an anti-ram structure is specified to reduce the likelihood of an engine block penetrating the defensive perimeter and sliding into the protected building. These dimensions are used for most tested barriers and have been accepted by the (U.S.) General Services Administration and the U.S. Department of State.

The performance of an anti-ram structure is judged with respect to a design or moving vehicle threat, which is usually a particular vehicle weight moving at a selected speed. Typical threats are a 4,000-lb (1,800-kg) car moving at 30 miles per hour (13 m/s) or a 15,000-lb (6,800-kg) truck moving at 50 miles per hour (22 m/s). Design speeds can be reduced below these values if the maximum achievable speed, resulting from the installation of barriers or other impediments, is lower.

Results of full-scale testing of a vehicle barrier may be extrapolated to barriers of similar construction by analysis and design through appropriate scaling of the demand (e.g., vehicle weight, impact speed, impact direction) and the resistance (e.g., mechanical properties, geometry, strength and deformation capacity). Extrapolation of test data must address differences, if any, in the foundation and soils supporting the anti-ram structure.

C2.3.2.2.3 Landforms. Landforms can be used to provide effective standoff against vehicular attack and to prevent line-of-sight attacks using shoulder-fired munitions.

C2.4 RISK ACCEPTANCE

Risk acceptance is an informed decision to tolerate the consequences and likelihood of a particular attack scenario. There are two ways in which risk acceptance can be expressed: design to risk and design to budget, as described below. The decision on the level of risk acceptance is often based on a risk-cost assessment that attempts to balance these two approaches. Consideration should also be given to allowing incremental risk reduction over a period of time in order to reach the defined risk acceptance level. In all cases, the owner must recognize and accept that it is certainly possible that the building could be subjected to blast effects in excess of those used for design, which may result in damage greater than anticipated.

Risk analysis in accordance with this chapter may lead to the selection by the building owner of a specific set of scenarios for which the facility will be explicitly designed. In such cases, it is usually not feasible to attempt to accommodate the maximum conceivable risk; instead, design is based on what is considered to be the maximum acceptable risk level. This is the design to risk approach.

The range of design options available to decision makers is extensive, as are the potential costs. Parallel to the reality of risk is the reality of budget constraints. Owners and managers of constructed facilities are confronted with the challenge of responding to the potential for an explosive event in a financially responsible manner. When sufficient funds are not available to design the building against the postulated threat, tradeoffs must be made. The level of protection that can be provided within the project budget then dictates the amount of risk that the owner must accept. This is the design to budget approach.

REFERENCES

American Petroleum Institute (API). (2003). Management of Hazards Associated with Location of Process Plant Buildings, API RP 752. API, Washington, D.C.

Federal Emergency Management Agency (FEMA). (2005). Risk Assessment: A How-To Guide to Mitigate Potential Terrorist Attacks Against Buildings, FEMA 452. FEMA, Washington, D.C.

FEMA. (2003a). Reference Manual to Mitigate Potential Terrorist Attacks Against Buildings, FEMA 426. FEMA, Washington, D.C.

FEMA. (2003b). Primer for Design of Commercial Buildings to Mitigate Terrorist Attacks, FEMA 427. FEMA, Washington, D.C.

National Fire Protection Association (NFPA). (2008). Flammable and Combustible Liquids Code, NFPA 30. NFPA, Quincy, Mass.

NFPA. (2007). Guide for Venting of Deflagrations, NFPA 68. NFPA, Quincy, Mass.

Sheet Metal and Air Conditioning Contractors' National Association (SMACNA). (1998). Seismic Restraint Manual. SMACNA, Chantilly, Va.

U.S. Department of Defense (DoD). (2008). Structures to Resist the Effects of Accidental Explosions, UFC 3-340-02, <www.wbdg.org/ccb/DOD/UFC/ufc_3_340_02.pdf> [May 12, 2011].

DoD. (2007a). DoD Minimum Antiterrorism Standards for Buildings, UFC 4-010-01, <http://www.wbdg.org/ccb/DOD/UFC/ufc_4_010_01.pdf> [May 12, 2011].

DoD. (2007b). DoD Security Engineering Facilities Planning Manual, UFC 4-020-01, <http://search.wbdg.org/wbdg/query.html?qt=UFC+4-020-01&charset=iso-8859-1&col=ccb> [May 12, 2011].

DoD. (1994). Security Engineering Final Design, TM 5-853-3, <http://search.wbdg.org/wbdg/query.html?col=wbdg&col=ccb&qt=TM+5-853-3> [May 12, 2011].

Chapter C3
PERFORMANCE CRITERIA

C3.2 DESIGN OBJECTIVES

C3.2.1 Limit Structural Collapse. The major cause of death and destruction when a building is subjected to the effects of a nearby explosion is collapse of all or a portion of the structure. Collapse can be mitigated by hardening key structural elements against a particular threat in conjunction with employing design and detailing practices that reduce the likelihood that a local failure will propagate through the structural system as a whole. Specific provisions that relate to the latter phenomenon are provided in Section 6.2.3.4.

Protective structural design generally follows two main principles: redundancy, which provides an alternative means of resisting the applied loads, especially for enhancing progressive collapse prevention; and target hardening, which involves performance improvements, such as strength and ductility, of individual elements and their interaction. The effectiveness of these measures depends on the characteristics of the building and of the expected blast load. Design and implementation should account for practicality and effectiveness in terms of costs—such as construction cost, relocation cost; and productivity loss—interruption to building operations, and intrusiveness to occupants. Priorities for vulnerability reduction should be given to the perimeter elements of the building.

A building's structural resistance is provided by integrating two load-carrying systems: the gravity load system and the lateral load system. A gravity load system is designed to carry the gravity load from the floors to the beams, to the girders, to the columns, and to the foundations. A lateral load system is designed to transfer horizontal loads from the superstructure to the foundations and to control the overall and interstory drift. The horizontal and vertical elements and their interfaces must be properly designed to withstand gravity and lateral loads and stress reversals.

Redundancy is related to both structural integrity and providing an adequate load path in the event of the failure of one element such that the structure remains stable. Redundancy usually requires ductility in individual elements and continuity in the structural system. Generally speaking, ductile moment frames with seismic detailing are typical of redundant systems, while unreinforced masonry walls and nonductile frames are representative of nonredundant systems.

The possibility of an explosive event larger than the design scenario should be recognized. The structure should be able to sustain local damage without destabilizing the whole structure. If local damage occurs under the assumed blast event, the structure should not collapse or be damaged to an extent disproportional to the original scope of the damage. For example, the failure of a primary structural element such as a beam, a slab, or a column should not result in failure of the structural system below, above, or in adjacent bays. In the case of column failure, damage to the beams and girders above the column should be limited to large inelastic deflections. Adequate redundancy and alternative load paths should be provided to reduce the vulnerability of a building against progressive collapse.

Performance considerations for vulnerability reduction in structural elements include adequate strength, ductility, and stiffness for each element and the interaction between connecting elements. Individual elements must meet the capacity and demand criteria for each element. Structural integrity and alternative load paths can be maintained through satisfactory interaction between connected elements, specifically proper connection detailing to ensure an adequate load transfer mechanism. However, reducing the vulnerability of individual elements by upgrading their strength and ductility characteristics is usually more effective than hardening only the connections.

In addition to considering the vulnerability reduction of structural elements, such as columns, walls, slabs, and beams, it is also desirable to consider vulnerability reduction for nonstructural elements such as non-load-bearing exterior and interior walls; window and framing systems; architectural, mechanical, electrical, and fire protection elements and systems; and these elements' interactions with the building structure.

Material design and detailing requirements vary from one construction material to another. Guidelines for concrete, steel, masonry, FRP, and other materials such as wood and cold-formed steel are available. Consideration of vulnerability reduction for existing buildings must take into account the compatibility of materials used and the impact on the structure's local and global response.

C3.2.2 Maintain Building Envelope. The envelope is the boundary between the exterior and interior of a building. Its primary function is to protect the controlled interior environment from the effects of uncontrolled exterior events, whether natural (such as temperature, rain, and wind) or human-induced (such as impact, forced entry, and blast). In the case of an exterior explosion, an envelope that remains intact will significantly reduce the hazards to people and property inside the building by preventing them from being subject to direct blast effects.

C3.2.3 Minimize Flying Debris. Where there is no building collapse, the major causes of injuries and damage are flying glass fragments and debris from walls, ceilings, and fixtures (nonstructural features). Flying debris can be minimized through building design and avoidance of certain building materials and construction techniques. The glass used in most windows breaks at very low blast pressures, resulting in hazardous, dagger-like shards. Minimizing those hazards through reduction in window numbers and sizes and through enhanced window construction has a major effect on limiting mass casualties. Window and door designs must treat glazing, frames, connections, and the structural elements to which they are attached as an integrated system. Hazardous fragments may also include secondary debris such as those from barriers and site furnishings.

Historically, overhead-mounted items are often dislodged in a blast event, posing a hazard to building occupants below. Because of this, the DoD requires the mountings for any such items that weigh more than 31 lb (14 kg) to be designed for a force equal to the component weight in the downward direction and 0.5 times the component weight in any other direction, including upward (DoD 2007a).

The vulnerability of facility occupants to all types of flying debris is largely a function of the building envelope design and construction. When the façade provides a higher level of protection, the likelihood of significant debris hazards on the interior of the structure is reduced.

C3.3 LEVELS OF PROTECTION

The performance goals specified in this Standard roughly correspond to those used in most building codes for natural hazards such as earthquakes. Higher LOPs are associated with more rigorous design and detailing requirements, and therefore higher construction costs, and should only be specified when justified by the results of a risk assessment in accordance with Section 2.2. The four LOPs designated as "Very Low," "Low," "Medium," and "High" derive from DoD criteria (DoD 2007a, 2007b) but are analogous to those used by other government and industry guidance documents. Since not all building occupants will necessarily survive an explosion, the performance goals primarily address the impact on surviving occupants.

The "Very Low" LOP should be considered a minimum baseline for buildings intended to provide meaningful resistance to blast effects. Structures that are not designed specifically to this level may fail catastrophically in the event of a nearby explosion, possibly leading to global or progressive collapse. Elements will be overwhelmed by the blast so that their debris is thrown with enough velocity to travel across the width of a typical room. This response realm can be useful when predicting the extent of severe injuries to room occupants.

C3.3.1 Structural Damage. Although this Standard is primarily intended for element-level analysis, realistic expectations for the overall damage to a building for each LOP are provided here for reference, consistent with DoD definitions (DoD 2007a, 2007b). Elements at each LOP do not necessarily cause the overall building to have the same LOP. Correlation between building LOP and element LOP is based in part on element type, for example, a cladding element versus a primary framing element (USACE 2008c).

C3.3.2 Element Damage. Element LOP definitions allow available blast test data to be categorized based on the observed response of actual construction assemblies and then used to help determine the corresponding response limits for analysis (USACE 2008c).

The acceptable LOP for a specific element should be determined based on the amount of protection the element must provide to building occupants and assets, and a consideration of how the response of that element affects the response of any attached elements. For example, first-floor load-bearing walls with a "Very Low" LOP can sustain enough damage to cause collapse of the roof or floors immediately above, and therefore affect the LOP of building occupants and assets over a much larger area of the building than a similar non-load-bearing wall (USACE 2008c).

This logic is also generally true, although to a lesser extent, for any failed roof framing elements, such as roof girders, that can cause failure of attached secondary elements, such as purlins or cladding. However, from a practical perspective, cladding elements and even secondary framing elements typically fail at blast loads much less than those needed to fail primary framing elements. The lower strength and failure of these elements often limits the blast load actually transferred into primary framing elements. Therefore, the response of cladding elements generally controls the LOP provided to building occupants and assets except in the case where a "Very Low" LOP occurs for a load-bearing wall or column, causing significant progressive collapse (USACE 2008c).

In light of the above considerations, the greater consequences of the failure of primary structural elements, as opposed to secondary structural elements and nonstructural elements, are addressed by assigning a lower level of acceptable damage to such elements for the same defined LOP (USACE 2008b).

Column response is of practical importance only for the case of column failure and associated progressive collapse, since lesser column damage may only potentially affect the very small percentage of building occupants and assets in the floor space directly behind the column (USACE 2008c).

C3.4 RESPONSE LIMITS

The maximum flexural response limits in Tables 3-2 and 3-3 have been developed by the U.S. Army Corps of Engineers Protective Design Center (PDC) by researching existing criteria and available test reports and consulting with technical experts in the field of blast effects analysis and design (USACE 2008b). These limits correspond to the four levels of protection defined in UFC 4-010-01 (DoD 2007a) and therefore are best suited for application to facilities required to resist terrorist explosive threats. Many of the limits are derived from a report prepared by Baker Engineering and Risk, Inc., for the *Component Explosive Damage Assessment Workbook* (*CEDAW*) software distributed by the PDC (USACE 2008c).

All of the limits are specifically intended for use with single-degree-of-freedom (SDOF) dynamic analysis of individual elements of conventional construction (not specifically detailed for blast resistance) subject to external far-range blast effects. They are not necessarily appropriate for other analysis approaches, internal blast loads with shock and gas pressure components, near-range blast effects, special performance expectations (e.g., explosive safety or direct weapons effects), or system-level evaluation. For example, the limits shown may not be applicable for complex finite element analyses. In such cases, performance limits based on fundamental principles of material and element constitutive models may be more appropriate.

C3.4.1 Flexural Elements. When using the SDOF methodology for flexural analysis and design of an element subject to blast effects, the following steps are usually followed (USACE 2008b):

- Select a trial member that is adequate to resist conventional loads.
- Determine the equivalent SDOF system.
- Calculate the airblast load based on the explosive quantity, standoff, and element orientation.
- Calculate the maximum dynamic deflection of the SDOF system.
- Calculate the corresponding ductility ratio and support rotation.
- Compare the response with established limits to determine the level of protection provided.
- Verify that failure in modes other than flexure, including flexural and direct shear, is precluded.
- Verify that failure will not occur due to load reversal or element rebound.

TABLE C3-1. RESPONSE PARAMETERS AND MODES

Element Type	Type of Response Parameter	Response Mode
Reinforced Concrete		
Slab (1-way or 2-way) or beam	Ductility ratio[a], support rotation	Elastic-perfectly-plastic flexural
Column (shear failure)	Ductility ratio	Shear
Masonry[b]		
Unreinforced	Ductility ratio[a], support rotation	Brittle elastic and arching
Reinforced	Ductility ratio[a], support rotation	Elastic-perfectly-plastic flexural
Structural Steel (Hot-rolled)		
Beam	Ductility ratio, support rotation	Elastic-perfectly-plastic flexural
Column (connection failure)	Ductility ratio	Connection shear
Column (flexural failure)	Ductility ratio, support rotation	Elastic-perfectly-plastic flexural
Open Web Steel Joist	Ductility ratio[a], support rotation	Elastic-perfectly-plastic flexural
Cold-Formed Steel		
Corrugated panel (1-way)	Ductility ratio, support rotation	Elastic-perfectly-plastic flexural
Girt or purlin	Ductility ratio, support rotation	Elastic-perfectly-plastic flexural
Stud connected at top and bottom	Ductility ratio, support rotation	Elastic-perfectly-plastic flexural
Stud with sliding connection at top	Ductility ratio, support rotation	Elastic-perfectly-plastic flexural
Wood Stud Wall	Ductility ratio, support rotation	Elastic-perfectly-plastic flexural

[a]Used only for "High" level of protection.
[b]Elements include brick, concrete masonry units, and European clay tile in single walls or cavity walls.
Source: USACE (2008c).

- Verify that all other design and detailing requirements are satisfied (Chapters 6 through 9).

Reversal and rebound are usually addressed for concrete elements by providing symmetric reinforcement. However, this is often not the case for prestressed concrete, in which case there is a risk of catastrophic, brittle failure when the element deflects in the direction opposite to that for which it is primarily designed.

It is important to note that the deflection and rotation limits specified for SDOF analysis do not necessarily correlate with the actual element deformations that would occur in a blast event; they take into account the assumptions and simplifications that come into play when modeling a continuous element as a lumped mass attached to a spring and, perhaps, a damper (USACE 2008c).

For the limits used in *CEDAW*, Table C3-1 provides the applicable nondimensional response parameter(s) and corresponding response mode for each element type. Ductility ratio and/or support rotation values for each level of protection were determined separately for each element type using trial and error so that the results were as consistent as possible with established guidelines (ASCE 1997, 1999; USACE 1994; DoD 2002, 2008) and available test data, consistent with the LOP definitions for element behavior in Section 3.3.2 (USACE 2008c).

In all cases except for columns, the "High" LOP response level was assumed to occur in flexural response at a ductility level of 1.0 based on the definitions in Section 3.3. For columns, only the "Low" level of protection was defined, corresponding to prevention of column failure. When more than one type of response parameter is applicable for a given LOP, the one that corresponds to the lowest blast loads governs (USACE 2008c).

Data for reinforced concrete slabs or beams are from blast testing of full-scale, half-scale, and quarter-scale one-way spanning reinforced concrete slabs with reinforcement ratios up to 0.66%. Single-reinforced elements with very high reinforcement ratios are less ductile in terms of support rotation. Therefore, the corresponding response limits are only applicable to cases where the reinforcement ratio does not exceed 0.5 times the value that produces balanced strain conditions. The resulting maximum acceptable reinforcement ratio for most of the test slabs would be approximately 1%. The quarter-scale tests were performed on slabs with equivalent full-scale dimensions of a 20-ft (6-m) span, 8-in. (200-mm) thickness, and 0.70% reinforcing ratio. Full-scale tests were performed on 8-ft (2.5-m) spanning walls with 6-in. (150-mm) and 8-in. (200-mm) thicknesses and reinforcing ratios of 0.20% and 0.25%. Half-scale tests were performed on a box-type concrete structure with 24-in. (600-mm), full-scale wall thickness and steel reinforcing ratios of 0.25% to 1.00% and shear reinforcement (USACE 2008c).

For prestressed concrete elements, partial prestressing should be considered when designing for blast loads, as mild steel can be used to minimize tensile cracking due to adverse displacements during member rebound. Prestressed concrete elements are normally considered to be very brittle for blast loads, and gravity-balanced members require additional top steel to control cracking due to dynamic uplift loads and displacements. Elements with $\omega_p > 0.30$ are over-reinforced and could fail catastrophically by concrete crushing before yielding occurs in the prestressing steel (USACE 2008b).

The higher response limits in Table C3-2 are commonly used for reinforced and prestressed concrete elements that have been properly designed and detailed to develop tension membrane behavior, which occurs at relatively large lateral displacements when the boundary conditions prevent inward, in-plane movement of the supports so that catenary action occurs. The element develops tension forces that form a couple with in-plane restraining forces at the supports to resist applied out-of-plane lateral loads. Theoretically, tension membrane response begins as soon as the element deflects if it has very rigid supports against in-plane movement. However, this level of support rigidity rarely occurs in buildings; consequently, tension membrane response is generally assumed to occur after flexural response. If the maximum tension force in the element is not limited by the connection strength, all of the stress in the maximum moment region transitions to tension stress at very large lateral deflections (USACE 2008a).

Tension membrane theory assumes that (1) all concrete has cracked throughout its depth and is incapable of carrying any

TABLE C3-2. FLEXURAL RESPONSE LIMITS FOR SDOF ANALYSIS—TENSION MEMBRANE[a]

Element Type	Expected Element Damage							
	Superficial		Moderate		Heavy		Hazardous	
	μ_{max}	θ_{max}	μ_{max}	θ_{max}	μ_{max}	θ_{max}	μ_{max}	θ_{max}
Reinforced Concrete								
Slab or beam (normal proportions[b])	1	–	–	6°	–	12°	–	20°
Slab or beam (deep elements[b])	0.6	–	–	3.6°	–	7.2°	–	12°
Prestressed Concrete								
Slab or beam[c] (normal proportions[b])	1	–	–	1°	–	6°	–	10°

[a]Where a dash (–) is shown, the corresponding parameter is not applicable as a response limit.
[b]Elements with normal proportions have a span-to-depth ratio ≥4; deep elements have a span-to-depth ratio < 4.
[c]Values assume bonded tendons, draped strands, and continuous slabs or beams.
μ, ductility ratio; θ, support rotation.
Source: USACE (2008b).

load, (2) all of the reinforcement has yielded and acts as a plastic membrane, and (3) only the reinforcement that extends over the entire area of the slab without splicing will contribute to the membrane. The theory further assumes that tension membrane action is dependent on the ductile behavior of the steel. It does not account for combined bending and tension membrane action. Experience has shown that some rupture of reinforcement may be expected in the tension membrane region. The rupture of some reinforcement is generally acceptable, since stirrups help maintain structural integrity by tying the reinforcement mats together and limiting the loss of damaged concrete. Therefore, for design purposes, 50% of the total principal reinforcement may be considered effective in the tension membrane region, consistent with available data. Also, full strain hardening will occur in the region of response and should be included.

Detailed information on the use of tension membrane in the design of reinforced concrete elements, as well as the requirements for ties or spirals in seismic columns, is available in the published literature (Park and Gamble 1999; DoD 2002).

Compression membrane response may occur after flexural response in an element that has rigid supports on both sides of the span length that do not allow rotation of the cross section. Compressive stresses develop in the unloaded face at supports and in the loaded face at the yield region near mid-span. The element can only rotate, and therefore deflect, as increasing amounts of crushing occur at these locations. Significant compression membrane resistance typically occurs only in relatively thick elements, such as reinforced concrete and masonry, rather than thin elements, such as steel plates and panels. In addition, compression membrane behavior is very sensitive to support flexibility and any small gaps between the responding element and those adjacent to it. Examples where it is commonly assumed to apply include infill concrete or masonry walls within a reinforced concrete frame, interior spans of a reinforced concrete slab that is continuous over its supports, and monolithic reinforced concrete box structures with heavy corner reinforcing steel (USACE 2008a).

Data for masonry are from the *Concrete Masonry Unit Database Software* (*CMUDS*). The full-scale equivalent dimensions for the data have spans of 8 to 11 ft (2.5 to 3.3 m) and thicknesses of 6 to 8 in. (150 to 200 mm). For reinforced masonry, the reinforcing ratios were 0.15% to 1.00%, although most of the data have reinforcing ratios from 0.15% to 0.30%. The data

include full-scale shock tube testing and high explosives testing on full- and quarter-scale walls (USACE 2008c). Elastic-perfectly-plastic analysis should not be used for unreinforced masonry elements, as this will lead to an unconservative calculation of the blast load that such elements can resist (USACE 2008b).

There are no available data for ductile flexural response of steel beams without tension membrane. All the available steel beam data are for light, cold-formed beams that were attached to steel framing. Response criteria for steel beams without tension membrane are available in published literature (ASCE 1997, 1999). The limits in Table 3-2 are derived primarily from these sources, but may not be appropriate for application to heavy steel girders (USACE 2008c). It is recommended that compact sections be used for all new designs (USACE 2008b).

Data for open web steel joists are from one blast test series on full-scale joists that supported a light metal deck roof system. The joists were 12K1 spanning 20 ft (6 m) at 4-ft (1.2-m) spacing. The joists were welded to steel plate embedded in the supporting reinforced concrete walls (USACE 2008c).

Data for cold-formed steel girts or purlins are from testing on full-scale, half-scale, and quarter-scale elements, where full-scale represents approximately an 8-in. (200-mm)-deep section with a 3-in. (75-mm) flange width and a material thickness in the range of 0.060 to 0.105 in. (16- to 12-gauge) with spans of 16 to 25 ft (5 to 7.5 m). Elements were assumed to have tension membrane equal to the lesser of the ultimate shear capacity of the bolts, the bearing capacity of the surrounding material, or the tensile capacity of the cross section. The data are from half-scale shock tube wall panel tests, quarter-scale tests conducted with high explosive on wall panels, several test series on whole pre-engineered buildings, and two test series where closely spaced, full-scale vertical members were bolted to a supporting frame to develop the full tensile capacity of the cross section and subjected to high explosive loads. In all but the last set of tests, the members supported typical corrugated steel panels (USACE 2008c).

There are some recent test data available for typical cold-formed steel stud walls subject to blast loads (DiPaolo and Woodson 2006). The steel studs are light elements, ranging in thickness up to 0.125 in. (3 mm), that can provide significant blast resistance when they are very well connected to their supports and can develop significant tension membrane capacity. The response limits in Table 3-2 are based on three different connection configurations: single or double slip tracks with screw attachment to the bottom track only; top and bottom channel tracks with screw attachments; and structural plate-and-angle connections at the top and bottom using a single bolt pivot (USACE 2008b).

Data for corrugated cold-formed steel panels are primarily from testing on two-span continuous full-scale panels ranging from light 24-gauge panels to heavy 3-in. (75-mm)-deep, 20-gauge panels with spans between 4 and 6 ft (1.2 and 1.8 m). Most of the tests were on corrugated steel panels attached to supporting members with self-tapping screws, but the data include several standing seam panels and insulated steel panels. Most of the data are from a test series with panels supported on rigid frames. The data also include shock tube testing of panels supported by lightweight girts and data from a test series on a full-scale pre-engineered building (USACE 2008c).

Most of the data for wood stud walls are from a series of tests on lightly constructed wood "SEA Huts" representing U.S. Air Force expeditionary structures loaded with large high-explosive charges at large standoffs. The walls had typical 2-in. × 4-in. (50 × 100 mm) wood studs at 16 in. (400 mm) on center, supporting

TABLE C3-3. FLEXURAL RESPONSE LIMITS FOR SDOF ANALYSIS—PETROCHEMICAL FACILITIES[a]

Element Type	Response Range					
	Low[b]		Medium[c]		High[d]	
	μ_{max}	θ_{max}	μ_{max}	θ_{max}	μ_{max}	θ_{max}
Reinforced Concrete						
Beams	–	1°	–	2°	–	4°
Slabs	–	2°	–	4°	–	8°
Beam-columns	1.3	1°	–	2°	–	4°
Reinforced Masonry						
1-way	1	0.5°	–	0.75°	–	1°
2-way	1	0.5°	–	1°	–	2°
Structural Steel						
Beams, girts, purlins	3	2°	10	6°	20	12°
Frame members[e]	1.5	1°	2	1.5°	3	2°
Plates	5	3°	10	6°	20	12°
Open Web Steel Joists	1	1°	2	1.5°	4	2°
Cold-Formed Steel Panels	1.75	1.25°	3	2°	6	4°

[a]Where a dash (–) is shown, the corresponding parameter is not applicable as a response limit.
[b]Localized building/component damage. Building can be used; however, repairs are required to restore integrity of structural envelope. Total cost of repairs is moderate.
[c]Widespread building/component damage. Building cannot be used until repaired. Total cost of repairs is significant.
[d]Building/component has lost structural integrity and may collapse due to environmental conditions (i.e., wind, snow, rain). Total cost of repairs approach replacement cost of building.
[e]Side-sway limits for frames: low = H/50, medium = H/35, high = H/25.
μ, ductility ratio; θ, support rotation.

TABLE C3-4. FLEXURAL RESPONSE LIMITS FOR SDOF ANALYSIS—PHYSICAL SECURITY[a]

Element Type	Damage Level		
	Light	Moderate	Severe
	θ_{max}	θ_{max}	θ_{max}
Reinforced concrete beams and slabs	2.3°	4.6°	8.5°
Structural steel beams	2.9°	6.8°	14°

[a]These values should not be used for design, but would be more suitable for postevent assessment.
θ, support rotation.
Source: ASCE (1999).

TABLE C3-5. FLEXURAL RESPONSE LIMITS FOR SDOF ANALYSIS—EXPLOSIVE SAFETY[a]

Element Type	Protection Category[b]			
	1[c]		2[d]	
	μ_{max}	θ_{max}	μ_{max}	θ_{max}
Reinforced Concrete				
Unlaced slabs with continuous edge support(s)	–	1°[e]	–	12°
Flat slabs	–	1°[e]	–	8°
Laced slabs with continuous edge support(s)	–	2°	–	12°
Beams with stirrups (closed ties)	–	2°	–	8°
Masonry[f]				
1-way	–	0.5°	–	1°
2-way	–	0.5°	–	2°
Structural Steel				
Beams, purlins, spandrels, girts	10	2°	20	12°
Frame structures	–	2°[g]	–	–
Plates	10	2°	20	12°
Open Web Steel Joists				
Controlled by maximum end reaction	1	1°	–	–
Otherwise	4	2°	–	–
Cold-Formed Steel Floor and Wall Panels				
Without tension membrane action	1.75	1.25°	–	–
With tension membrane action	6	4°	–	–

[a]Where a dash (–) is shown, the corresponding parameter is not applicable as a response limit.
[b]U.S. Department of Defense (DoD 2008) also defines response limits for two lower levels of protection. Protection Category 3 is used to prevent the propagation of an explosives detonation from one area containing explosives to another. Protection Category 4 is used to prevent a prompt propagation between such areas.
[c]Protect personnel against the uncontrolled release of hazardous materials, including toxic chemicals, active radiological and/or biological materials; attenuate blast pressures and structural motion to a level consistent with personnel tolerances; and shield personnel from primary and secondary fragments and falling portions of the structure and/or equipment.
[d]Protect equipment, supplies, and stored explosives from fragment impact, blast pressures, and structural response.
[e]2 deg with either shear reinforcement or tensile membrane action.
[f]Response limits for Protection Categories 1 and 2 are for reusable and non-reusable conditions, respectively.
[g]Relative side-sway deflection between stories is limited to H/25.
μ, ductility ratio; θ, support rotation.

5/8-in. (16-mm) plywood panels. Data from walls subject to side-on loads were not used if the roof failed or had severe damage, since the roof helped support these walls. There are also some data from shock tube tests on heavy wood stud walls, where 2-in. × 6-in. (50 × 150 mm) and 8-in. (200 mm) studs were spaced at 6 in. (150 mm) on center, supporting 5/8-in. (16-mm) plywood on one side or both sides of the wall, spanning 8 ft (2.5 m). In some of these tests the plywood was nailed to the studs with nail spacing as close as 3 in. (75 mm), but the peak measured dynamic reaction forces were consistent with reactions based only on the maximum resistance of studs with no composite action from the plywood (USACE 2008c).

Response limits for glazing system framing elements are taken from draft DoD guidance.

Tables C3-3, C3-4, and C3-5 provide alternative flexural response limits for SDOF analysis. The references from which they are taken explain the basis for them, including any associated design and detailing considerations—which may be significantly more restrictive than conventional construction—and they should only be used accordingly.

C3.4.2 Compression Elements. In SDOF analysis, deflection is determined for only one mode of response, typically flexure about a single axis. Exterior walls and columns subject to lateral blast loads are frequently also required to carry vertical loads, either conventional or blast. Limiting the deflections of these elements is vital in avoiding localized collapse. However, there is insufficient information available to develop response limits specifically for elements in combined flexure and compression. In addition, much of the compressive load carried by such an element is from the reactions of supported elements loaded by the blast; determination of the magnitude and timing of these reactions is difficult at best. Consequently, more conservative

limits are placed on the flexural response of an element when the amount of axial compressive load exceeds 10% of its ultimate dynamic axial compression capacity. In general, the response limits are those that correspond to a moderate damage level.

Available data from typical reinforced concrete buildings indicate that the columns are much more resistant to blast loads than are the surrounding wall cladding elements. Typically, the cladding in reinforced concrete frame buildings spans vertically between floors and does not transfer blast load onto the columns, so that the columns are only loaded over their own width. Also, the columns are required to have a significant percentage of longitudinal steel to resist conservatively high, design-level, axial loads, and this steel acts as flexural reinforcement under lateral blast loads to provide a very significant moment-resisting capacity. This is true even when considering P-delta effects from axial loads because the columns are also very stiff laterally and therefore tend to have small lateral deflections. Except in earthquake zones, the only lateral steel reinforcement in columns are typically column ties that are too widely spaced to provide significant shear strength. Even in columns with closely spaced stirrups, the lateral load capacity is almost always controlled by the column shear strength instead of flexural strength (USACE 2008c).

The same rationale also generally applies to structural steel columns, specifically perimeter ground-level columns where the connections are in shear—typically at the bottom where there is a shear plane through the anchor bolts connecting the column bearing plate to the concrete slab. The cladding does transfer blast loads into columns for some structural steel frame types, such as pre-engineered buildings, but the cladding typically has a much lower blast capacity than the columns in these cases and fails before the frame members. Of course, when cladding is expected to transfer blast loads to columns, the columns must be designed to transfer these loads to lateral resistance systems.

Limited available blast test data for structural steel columns subject to severe blast loads indicate that connection failure is the weakest response mode when conventional types of base plate connections are used (USACE 2008c). Because of this, it is recommended that the perimeter base plates in structural steel frame buildings be buried in the concrete slab or continuous into a basement so that the column bears against the ground-floor slab (USACE 2008b). If the column is also continuous over the second-floor slab, so that the ground-floor column capacity is not controlled by connection capacity, flexure is expected to control column response to close-in blast loads. The columns can be considered vertical beams loaded over their flange widths, or the width exposed to blast (USACE 2008c).

C3.5 ELEMENT STRENGTH

C3.5.1 Strength Increase Factors. Designers typically specify a minimum value for each property that contributes to the strength of an element. Consequently, the actual strength of material supplied for use in construction will almost always exceed this value. The average strength factor (ASF) is intended to account for this by assuming an expected actual strength that is greater than the specified minimum value, typically by 10% for concrete and steel (ASCE 1997; USACE 1994; DoD 2008). In the case of cold-formed steel, an actual ASF of 1.21 is combined with a reduction factor on the plastic moment capacity of 0.9 to account for the unique nonlinear behavior of such sections, resulting in an effective ASF of 1.1 (DoD 2008).

The compressive strength of moist-cured concrete will continue to increase with age beyond the 28 days normally specified. This is often incorporated into analysis and design by applying an age increase factor, which for concrete made from Type I portland cement is equal to 1.10 at 6 months and 1.15 at 1 year. The corresponding values for high-early-strength concrete made from Type III Portland cement are 1.05 and 1.08 (USACE 1994).

In addition, blast loads tend to have very high magnitudes and very short durations. As a result, they develop very high strain rates in responding elements that cause an effective increase in the compressive strength of concrete or masonry and the yield

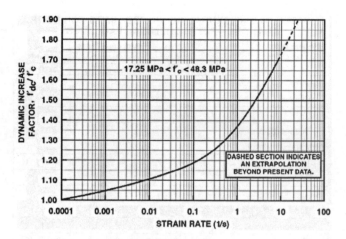

FIGURE C3-1. DIF FOR CONCRETE COMPRESSIVE STRENGTH (USACE 2008a).

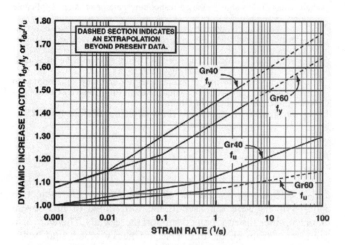

FIGURE C3-2. DIF FOR REINFORCING STEEL YIELD AND ULTIMATE STRENGTH (USACE 2008a).

strength of steel. The dynamic increase factors (DIF) in Table 3-5 are intended to account for this at particular strain rate values that are conservative in most cases (ASCE 1997; DoD 2008). However, they should be used with caution when evaluating the effects of a long-duration blast load, such as that resulting from a vapor cloud explosion. Figures C3-1, C3-2, and C3-3 provide alternative flexural DIF values for concrete, reinforcing steel, and structural steel over a range of strain rates (USACE 2008a).

Figure C3-4 illustrates conceptually how the specified minimum strength of an element is adjusted by the ASF and DIF, using a histogram of hypothetical test results. This procedure should be used with caution for existing components whose condition is questionable, or whose exact construction is not well defined.

C3.5.2 Strength Reduction Factors. Codes and specifications for strength design of structural elements typically require $\phi < 1.0$ so that element strengths are reduced below nominal values to account for the following:

- Epistemic uncertainty in analysis, e.g., computation of demands on elements
- Aleatory randomness in material properties
- Epistemic uncertainty in the computation of some member strengths, e.g., shear strength of concrete

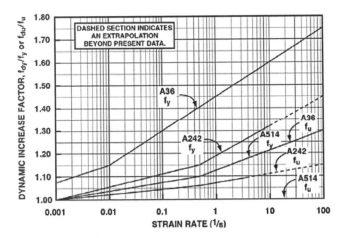

FIGURE C3-3. DIF FOR STRUCTURAL STEEL YIELD AND ULTIMATE STRENGTH (USACE 2008a).

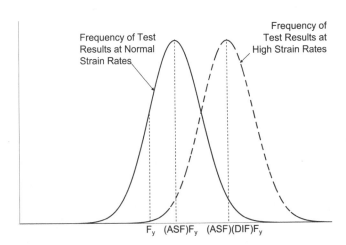

FIGURE C3-4. EFFECT OF STRENGTH INCREASE FACTORS.

* The mode of failure, e.g., ductile or brittle, at the element and system levels.

The variability in the design equations for element resistance is greater for shear and axial strength than for flexure; this is addressed by requiring smaller values of ϕ.

Published values of ϕ are based primarily on experience with constant or steadily increasing applied loads on structural elements that are intended to remain elastic. They are calibrated in conjunction with specific load combinations to provide a certain level of reliability consistent with historic practice using allowable stress design and unfactored service loads. However, blast loads typically deform elements well beyond the elastic limit, at which point the randomness of the material properties is relatively insignificant. Structures designed in this regime are controlled by the duration of the loading, the (actual) resistance of the element, and the configuration of the boundary conditions. Proper analysis for blast effects requires closely determining an average expected deformation of the structure or element being analyzed. Safety is addressed in the selection of appropriate ASFs, DIFs, and response limits; therefore, it is common practice to set $\phi = 1.0$, especially with respect to flexure.

The ASFs, DIFs, and flexural response limits specified in this Standard for SDOF analysis are all based on $\phi = 1.0$. The assessment of existing elements, as well as all other cases that involve

estimating the actual response, should utilize $\phi = 1.0$, unless a different value can be justified in a specific situation. For the design of new elements, it is acceptable and generally conservative to use the common values of $\phi < 1.0$. However, there are exceptions; for example, when verifying that shear and connection capacities are adequate to develop the full flexural capacity of an element, the shear and connection capacities can certainly be calculated using $\phi < 1.0$, but the flexural capacity should be calculated using $\phi = 1.0$ to avoid underestimating the corresponding shear and connection demand.

C3.5.3 Remaining Strength. At least some gravity and environmental loads are likely to be acting on an element when a nearby explosion occurs, even though their effects may not be of the same nature or in the same direction. Codes and specifications for strength design of structural elements typically include equations that address the interaction of the various types of loads that may be applied, such as axial, flexural, torsional, and shear effects. Equations 3-1 and 3-2 are derived from the 2005 Commentary of ASCE/SEI 7 (ASCE 2005) and are intended for checking the capacity of a structure or structural element to withstand the effect of an extraordinary event, such as a fire, explosion, or vehicular impact. Equation 3-1 has been moved to the mandatory portion of the 2010 version of ASCE/SEI 7 (ASCE 2010), while Eq. 3-2 has been eliminated.

The uncertainty of the blast load is encompassed in the selection of its magnitude, and thus its load factor is usually set equal to 1.0. The dead load factor of 1.2 when the dead load reduces the strength available to resist the blast load (e.g., lateral force on a column), or 0.9 when the dead load increases the strength available to resist the blast load (e.g., uplift on a floor slab), is intended to account conservatively for uncertainty during design in the estimation of the weight of all materials of construction that will be incorporated into the building. For existing buildings, this uncertainty is often minimal, so that a load factor of 1.0 is usually appropriate. Load factors less than 1.0 on the companion actions reflect the small probability, in most cases, of a joint occurrence of the blast load and the design live, snow, or wind load. The default companion action $0.5L$ corresponds, approximately, to the mean of the yearly maximum; however, in situations where an explosion is likely to occur during a design live load event, such as a terrorist attack against a highly occupied facility, then $1.0L$ should be substituted. Companion actions $0.2S$ and $0.2W$ are interpreted similarly.

The term $0.2W$ in Eq. 3-2 is intended to ensure that the lateral stability of the structure is checked. Some recent standards require the stability of the structure to be checked by imposing a small notional lateral force equal in magnitude to 0.2% of the total gravity force due to the summation of the dead and live loads acting on the story above that level. If such a stability check is performed, Eq. 3-2 need not be considered.

Increased mass generally has a beneficial effect on the response of an element to blast loads. Consequently, it is important not to overestimate the mass that is providing inertial resistance to deformation of the element, including the mass of the element itself and any attached construction. Equations 3-1 and 3-2 are intended to apply to stress calculations only, not to mass calculations.

REFERENCES

American Society of Civil Engineers (ASCE). (2005). Minimum Design Loads for Buildings and Other Structures, ASCE/SEI 7-05. ASCE, Reston, Va.

American Society of Civil Engineers (ASCE). (2010). Minimum Design Loads for Buildings and Other Structures, ASCE/SEI 7-10. ASCE, Reston, Va.

American Society of Civil Engineers, Task Committee on Structural Design for Physical Security of ASCE (ASCE). (1999). Structural Design for Physical Security: State of the Practice. ASCE, Reston, Va.

American Society of Civil Engineers, Task Committee on Blast-Resistant Design of the Petrochemical Committee of the Energy Division of ASCE (ASCE). (1997). Design of Blast Resistant Buildings in Petrochemical Facilities. ASCE, Reston, Va.

DiPaolo, B. P., and Woodson, S. C. (2006). "An overview of research at ERDC on steel stud exterior wall systems subjected to severe blast loading." *Proc., ASCE/SEI Structures Congress, 2006.* ASCE, Reston, Va.

Park, R., and Gamble, W. L. (1999). *Reinforced concrete slabs*, 2nd ed. John Wiley & Sons, New York.

Salim, H., Dinan, R., and Townsend, P. T., "Analysis and experimental evaluation of in-fill steel stud wall systems under blast loading." *ASCE J. Struct. Eng.*, 131(8), 1216–1225.

U.S. Army Corps of Engineers (USACE). (2008a). Methodology Manual for Single-Degree-of-Freedom Blast Effects Design Spreadsheets (SBEDS), PDC-TR 06-01. Protective Design Center, USACE, Omaha, Neb.

USACE. (2008b). Single Degree of Freedom Response Limits for Antiterrorism Design, PDC-TR 06-08. Protective Design Center, USACE, Omaha, Neb.

USACE. (2008c). Methodology Manual for Component Explosive Damage Assessment Workbook (CEDAW), PDC-TR 08-07. Protective Design Center, USACE, Omaha, Neb.

USACE. (1994). Facility and Component Explosive Damage Assessment Program (FACEDAP) Theory Manual, TR 92-2. Protective Design Center, USACE, Omaha, Neb.

U.S. Department of Defense (DoD). (2008). Structures to Resist the Effects of Accidental Explosions, UFC 3-340-02, <www.wbdg.org/ccb/DOD/UFC/ufc_3_340_02.pdf> [May 12, 2011].

DoD. (2007a). DoD Minimum Antiterrorism Standards for Buildings, UFC 4-010-01, <http://www.wbdg.org/ccb/DOD/UFC/ufc_4_010_01.pdf> [May 12, 2011].

DoD. (2007b). DoD Security Engineering Facilities Planning Manual, UFC 4-020-01, <http://www.wbdg.org/ccb/DOD/UFC/ufc_4_020_01.pdf> [May 12, 2011].

DoD. (2002). Design and Analysis of Hardened Structures to Conventional Weapons Effects, UFC 3-340-01 (for official use only).

Chapter C4
BLAST LOADS

C4.1 GENERAL

This version of the Standard contains simple procedures for commonly encountered external and internal blasts. The Standard requires the use of recognized literature in other cases and permits the same in all. This commentary includes guidance on this body of information. This chapter is presented in the units of measure that are commonly used in U.S. practice.

C4.2 BASIC PROCEDURE FOR EXTERNAL BLAST

C4.2.1 Scope. Surrounding structures and terrain can have a shielding or focusing effect on the blast loading. In the former case the basic procedure represents a conservative approach. The designer should consider the possibility of focusing in the application of this procedure and may consider shielding as appropriate.

The shape of the structure can affect the load applied from a blast. The basic procedure applies to those that are regular in their spatial form. It does not apply to those having a clearly unusual geometry, such as a reentrant corner that may focus the blast.

C4.2.2 Directly Loaded Surfaces. The basic procedure for directly loaded surfaces is taken from the DoD practice for hardened structures (UFC 3-340-01; DoD 2002). Any wall facing the source of a blast, i.e., the front wall, is directly loaded by the reflection of the incident and dynamic pressures. An equivalent triangular load function is first generated for the positive phase of the normal reflected shockwave without clearing. Then a bilinear load function is generated for the positive phase of the normal reflected shockwave with clearing. This function changes slope at time t_c where, due to clearing, the reflected overpressure decays to the same level as the side-on overpressure. Since the impulse derived using the triangular function may be unnecessarily conservative in some cases, the positive phase impulse of the normal reflected shockwave is taken as the lesser of that given by the equivalent triangular function and the bilinear function. A similar approach is used for oblique reflected shockwaves, but in that case the peak reflected overpressure is modified to account for the effects of the angle of incidence. The negative phase loading, if needed, can be approximated for normal and oblique reflected shockwaves by an equivalent triangular function as described in Section 4.2.2.

Often the design threat is expressed as an equivalent mass of trinitrotoluene (TNT), W_e. Table 4-1 provides the factors to convert quantities of other high explosives into their TNT equivalents. In general, this effective mass is different for pressure and impulse. Table 4-1 gives averaged values for several explosives. For compounds not tabulated, the equivalence can be approximated by the ratio of the heats of detonation. For other than high explosive events, such as vapor cloud deflagration or pressure vessel burst, specialized procedures must be used to estimate W_e.

Figures 4-3 through 4-8 are based in large part on empirical fits to experiments. Thus, they should not be extrapolated beyond the ranges shown. The nonmonotonic character of the reflected overpressure coefficients in Fig. 4-6 has been questioned and is currently under study by DoD.

C4.2.3 Indirectly Loaded Surfaces. The basic procedure for indirectly loaded surfaces is taken from the DoD practice for hardened structures (UFC 3-340-01; DoD 2002). Any wall not facing the source of a blast, i.e., the roof, the side walls, and rear wall, is loaded by the incident overpressure and a negative dynamic pressure. Figure 4-1 idealizes this loading as a linear decay. The parameters are selected to preserve the peak overpressure and the total impulse of the actual loading. The loading function is based on the peak side-on overpressure combined with the drag overpressure calculated using the appropriate values from Table 4-2.

C4.3 BASIC PROCEDURE FOR INTERNAL BLAST

In an internal explosion each surface is subjected to shock waves from the reflections off multiple surfaces. Internal explosions also produce gaseous products and associated overpressures whose magnitude is generally less than that of the shock overpressure but whose duration is significantly longer. This complex loading is comprehensively addressed in the DoD procedures for design to resist accidental explosions (UFC 3-340-02; DoD 2008).

The basic procedure conservatively applies the gas overpressure as a static load on the affected surfaces. In some situations this will result in an acceptable design. In others the design may be excessively conservative due to venting and the comprehensive methods of UFC 3-340-02 (DoD 2008) can be used.

The basic procedure was developed for the SDOF analysis of structural elements that may be subjected to far range blast loading only (scaled distance $Z \geq 3.0$). This limit is consistent with the SDOF response and analysis limits of Chapters 3 and 6. For close-in loading, a more rigorous shock and gas pressure calculation procedure, such as that provided in UFC 3-340-02, should be applied.

The quantitative limits on the vent area and the unit weight of the vent(s) are based on a series of parametric calculations in which the results of UFC 3-340-02 were compared to the basic procedure.

In Figs. 4-9 and 4-10 for the peak gas overpressure, the *free* volume of the interior, i.e., the total volume minus the volume of all interior equipment, structural elements, etc., should be used in calculating the loading density. As these figures are based on empirical fits to experiments, the curves should not be extrapolated beyond the range shown.

C4.4 OTHER PROCEDURES

Analytical Methods. Two general types of analytical methods have been developed to predict blast loads. The semi-empirical approach employs a physics based model to compute blast parameters with coefficients that are selected to match test data. Such methods are relatively convenient to use but are limited to the conditions for which they have been calibrated. A number of these are discussed further in (ASCE 1999). The hydrodynamic approach solves the underlying thermodynamic and hydrodynamic equations using numerical analysis. These models are more versatile but require some judgment in application.

Experimental Methods. Experimentation is sometimes an attractive approach for particularly complex and critical structures. This is often economically performed at reduced geometric scale and extrapolated to prototype conditions using dimensional analysis.

REFERENCES

American Society of Civil Engineers, Task Committee on Structural Design for Physical Security of ASCE (ASCE). (1999). Structural Design for Physical Security: State of the Practice. ASCE, Reston, Va.

U.S. Department of Defense (DoD). (2008). Structures to Resist the Effects of Accidental Explosions, UFC 3-340-02, <www.wbdg.org/ccb/DOD/UFC/ufc_3_340_02.pdf> [May 12, 2011].

DoD. (2002). Design and Analysis of Hardened Structures to Conventional Weapons Effects, UFC 3-340-01 (for official use only).

Chapter C5
FRAGMENTATION

C5.1 GENERAL

C5.1.1 Scope. Design for fragmentation is typically outside standard practice for commercial buildings. Two types of fragments may be generated during an explosion: primary and secondary fragments. The effects of fragments may be considered on the exterior envelope or specific locations within the building (e.g., safe havens). This chapter provides analytical methods for estimating the hazards of secondary fragmentation only.

Symbols and Notation

The following symbols and notation apply only to the provisions of the Commentary of Chapter 5:

A = projected or mean presented area of object, in.^{2} (mm^2)

A_b = beam cross-sectional area perpendicular to long direction, in.^{2} (mm^2)

a = fragment radius, assuming spherical shape, in. (mm)

a_o = velocity of sound in air, 1,100 ft/s (335 m/s)

b = loaded beam width, ft (m)

C_d = drag coefficient; see Table C5-1

C_s = dilatational velocity of elastic wave through concrete, ft/s (m/s)

C_0 = constant in TNT equivalent, equal to 6,950 lbf-s/ft^3 (1.091×10^6 N-s/m^3)

C_1 = constant, equal to −0.2369 for cantilever beam, −0.6498 for clamped-clamped beam

C_2 = constant, equal to 0.3931 for cantilever beam, 0.4358 for clamped-clamped beam

C_3 = constant; see Table C5-2

d = equivalent fragment diameter, in. (mm)

E_c = concrete modulus of elasticity, psi (N/m^2)

f'_c = compressive strength of concrete target, psi (N/m^2)

g = acceleration of gravity equal to 32.2 ft/sec^2 (9.81 m/s^2)

H = the minimum transverse distance of the mean presented area, in. (mm)

h = target thickness, in. (mm)

i = specific acquired impulse, lbf-s/ft^2 (N-s/m^2)

$\bar{\imath}$ = nondimensional specific impulse

i_s = specific impulse, psi-sec (Pa-s)

K = constant; see Table C5-3

L = loaded beam length, ft (m)

m_f = mass of fragment, lbm (kg)

$\bar{p}$ = nondimensional pressure

p_o = atmospheric pressure, equal to 14.70 psi (1.013×10^5 Pa)

p_{so} = peak incident overpressure, psi (Pa)

R = distance from explosive charge center to nearest surface of object (standoff distance); see Fig. C5-1, ft (m)

$\bar{R}$ = nondimensional fragment range

R_e = radius of explosive, ft (m)

R_{eff} = effective radius of the equivalent sphere of explosive ($R_{eff} = R_e$ for a spherical charge), ft (m)

R_{max} = maximum fragment range, ft (m)

R_t = radius of target; see Fig. C5-1, in. (mm)

T = fragment toughness; see Table C5-4, in.-lbf/in.^{3} (mm-N/mm^3)

T_{pf} = minimum thickness required to prevent perforation, in. (mm)

T_{sp} = minimum thickness required to prevent spalling, in. (mm)

$\bar{v}$ = nondimensional object velocity

v_o = initial fragment velocity, ft/sec (m/s)

v_{st} = fragment striking velocity, ft/sec (m/s)

v_{50} = limit velocity, ft/sec (m/s)

$\overline{v_{50}}$ = nondimensional limit velocity

w_f = fragment weight, lbf (N)

X = distance from the front of the fragment to the largest cross-sectional area, in. (mm)

x_f = maximum penetration for steel fragment through concrete wall, in. (mm)

α_o = initial trajectory angle, degrees (radians)

β = object shape factor; see Fig. C5-1

ρ_o = air mass density, equal to 2.378×10^{-3} lb-s^2/ft^4 (0.125 kg-s^2/m^4)

ρ_f = fragment mass density, lb/ft^3 (kg/m^3)

ρ_t = target mass density, lb/ft^3 (kg/m^3)

σ_t = target yield stress, lbf/ft^2 (N/m^2)

C5.2 DESIGN REQUIREMENTS

C5.2.1 Design Requirements.

Fragmentation Failure. UFC 3-340-02 (DoD 2008) defines direct spalling as the dynamic disengagement of the surface of an element, typically concrete, resulting from a tension failure normal to its free surface caused by shock pressures of an impinging blast wave being transmitted through the element. The same document defines "scabbing" as the dynamic disengagement of the surface normal to its free surface caused by large strains in the flexural reinforcement.

C5.3 ANALYTICAL PROCEDURES

Multiple methods are acceptable for fragmentation analysis, including the methods presented herein. Other acceptable methods, including for materials not covered in this section, may be found in the references of this chapter (DoD 2008; DoE 1992; Baker et al. 1983).

C5.3.1 Acceptable Analytical Methods. The general methodology for identifying the hazard associated with secondary fragmentation is listed below:

1. Define the explosive and fragment threat.
2. Calculate the initial velocity of the fragment.

TABLE C5-1. DRAG COEFFICIENTS (Cd) FOR VARIOUS SHA

Shape	Sketch	C_D
Circular cylinder (Long rod), side-on	FLOW	1.20
Sphere	FLOW	0.47
Rod, end-on	FLOW	0.82
Disc, face-on	FLOW OR	1.17
Cube, face-on	FLOW	1.05
Cube, edge-on	FLOW	0.80
Long rectangular member, face-on	FLOW	2.05
Long rectangular member, edge-on	FLOW	1.55
Narrow strip, face-on	FLOW	1.98

Source: DOE (1992).

TABLE C5-2. Fragment Penetration Factors, C_3

Type of Material	C_3
Armor-piercing steel	1.00
Mild steel	0.70
Lead	0.50
Aluminum	0.25

Source: DOE (1992).

TABLE C5-3. CONSTANT, K

Description	K Value
Object on reflecting surface (i.e., ground)	4
Object in air	2

Source: DOE (1992).

3. Calculate the fragment trajectory to determine the striking velocity at the impact location.
4. Estimate the fragment impact damage.

Initial Velocity of Fragment

Unconstrained Secondary Fragments Far from Charge. For objects located at a distance more than 20 times the radius of explosive, the object is considered "far away" and the initial fragment velocity should be estimated using this section (DoE 1992).

1. The peak incident overpressure (p_{so}) and incident specific impulse (i_s) are calculated per Chapter 4.
2. The following parameters are determined or assumed for the size, shape, and material of typical fragments: mass of fragment (m_f), drag coefficient (C_d) (see Table C5-1), distance from the front of the fragment to the largest cross-sectional area (X), minimum transverse distance of the mean presented area (H), and mean presented area (A)
3. From the location of the object, determine the constant, K, from Table C5-3.
4. Calculate the nondimensional specific impulse ($\bar{i}$) and nondimensional pressure ($\bar{p}$), then determine the nondimensional object velocity ($\bar{v}$) from Fig. C5-2.

$$\bar{i} = \frac{C_d \cdot i_s \cdot a_o}{p_{so} \cdot (K \cdot H + X)} \quad \text{(C5-1)}$$

$$\bar{p} = \frac{p_{so}}{p_o} \quad \text{(C5-2)}$$

5. The initial fragment velocity (v_o) is calculated from the nondimensional object velocity:

$$v_o = \bar{v} \cdot \left[\frac{p_o \cdot A \cdot (K \cdot H + X)}{m_f \cdot a_o} \right] \quad \text{(C5-3)}$$

Unconstrained Secondary Fragments Close to Charge. For objects located at a distance less than or equal to 20 times the radius of explosive, the object is considered "close" to the charge. The initial fragment velocity should be estimated using this section for spherical charges (DoE 1992).

The specific acquired impulse (i) is determined as follows:

1. If $R/R_e < 5$, i is determined from equations below.
2. If $5 < R/R_e \leq 20$, i equals the normally reflected impulse (see Chapter 4).
3. The target or object shape factor (β) is given in Fig. C5-1.
4. The following parameters are determined from the size and shape of typical fragments: mean presented area (A) and fragment mass (m_f).
5. Calculate the effective radius of a spherical charge:

$$R_{eff} = R_e \quad \text{(C5-4)}$$

6. For $R/R_e \leq 5.0$, then the specific acquired impulse (i) for TNT equivalent is:

$$i = (C_0 \cdot \beta \cdot R_{eff}) \cdot \left(\frac{R_e}{R} \right)^{1.4} \cdot \left(\frac{R_e}{R_t} \right)^{-0.158} \quad \text{(C5-5)}$$

7. The initial fragment velocity (v_o) is calculated as follows:

$$v_o = \frac{A \cdot \beta \cdot i}{m_f} \quad \text{(C5-6)}$$

Constrained Secondary Fragments. Estimation of constrained secondary fragments assumes that the specific impulse used to throw the object equals the total specific impulse less the impulse

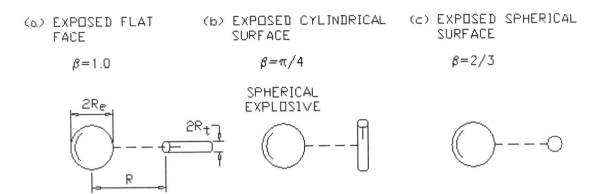

(a) EXPOSED FLAT FACE

$\beta = 1.0$

(b) EXPOSED CYLINDRICAL SURFACE

$\beta = \pi/4$

(c) EXPOSED SPHERICAL SURFACE

$\beta = 2/3$

FIGURE C5-1. TARGET SHAPE FACTOR (β) FOR UNCONSTRAINED FRAGMENTS.

TABLE C5-4. FRAGMENT TOUGHNESS, *T*

Steel	Toughness (in.-lbf/in^3) [mm-N/mm^3]
ASTM A36	12,000 [83]
ASTM A441	15,000 [103]
ASTM A514 Grade F	19,000 [131]

Source: DOE (1992).

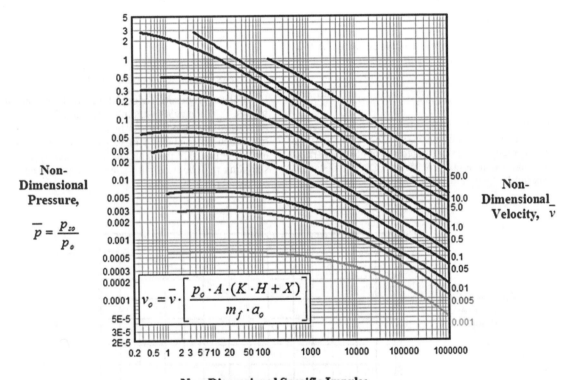

Non-Dimensional Pressure,

$$\overline{p} = \frac{p_{so}}{p_o}$$

Non-Dimensional Velocity, $\overline{v}$

$$v_o = \overline{v} \cdot \left[\frac{p_o \cdot A \cdot (K \cdot H + X)}{m_f \cdot a_o} \right]$$

Non-Dimensional Specific Impulse,

$$\overline{i} = \frac{C_d \cdot i_s \cdot a_o}{p_{so} \cdot (K \cdot H + X)}$$

FIGURE C5-2. NONDIMENSIONAL OBJECT VELOCITY AS A FUNCTION OF NONDIMENSIONAL PRESSURE AND NONDIMENSIONAL IMPULSE.

required to free the object from its moorings. The initial fragment velocity for such fragments from a cantilever or clamped-clamped beam should be estimated using this section (DoE 1992).

1. The following parameters are determined or estimated from the size, shape, and material of typical fragments: beam mass density (ρ_f), loaded beam length (L), loaded

beam width (b), beam cross-sectional area (A_b), and fragment toughness (T) (see Table C5-4).

2. The specific acquired impulse (i) is determined using Chapter 4.

3. The initial velocity after break-away (v_o) is calculated as follows, where C_1 and C_2 are constants dependent on end conditions:

If $\dfrac{i \cdot b}{\sqrt{\rho_f \cdot T \cdot A_b}} \cdot \left(\dfrac{L}{b/2}\right)^{0.3} \leq 0.602,$ (C5-7)

then $v_o = 0 \cdot ft/s$ [m/s]   (C5-8)

Otherwise, $v_o = \left[C_1 + C_2 \cdot \left(\dfrac{i \cdot b}{\sqrt{\rho_f \cdot T \cdot A_b}}\right) \cdot \left(\dfrac{L}{b/2}\right)^{0.3} \right] \cdot \left(\sqrt{T/\rho_f}\right)$

    (C5-9)

Fragment Trajectory

"Chunky" or Drag-Controlled Fragments. The maximum fragment range (R_f) should be estimated using this section (DoE 1992).

1. The following parameters are determined or estimated from typical fragments: mass density of air (ρ_o), drag coefficient (C_d) (see Table C5-1), fragment drag area (A) fragment initial velocity (v_o), and fragment mass (m_f).
2. The nondimensional velocity ($\bar{v}$) is calculated as follows:

$$\bar{v} = \dfrac{\rho_o \cdot C_d \cdot A \cdot v_o^2}{m_f} \quad (C5\text{-}10)$$

3. The nondimensional range ($\bar{R}$) is determined from the ordinance of Fig. C5-3.
4. The maximum fragment range is calculated using the following equation:

$$R_{\max} = \dfrac{m_f \cdot \bar{R}}{\rho_f \cdot C_d \cdot A} \quad (C5\text{-}11)$$

Basic Kinematic Equations. Basic kinematic equations are acceptable for estimating the fragment striking velocity (v_{st}), initial fragment velocity (v_o), initial trajectory angle (α_o), fragment mass (m_f), and fragment drag coefficient (C_d), and distance to the target should be taken into account.

Fragment Impact Damage. The fragment impact damage may be estimated using this section (DoE 1992).

Limit Velocity of "Chunky" Nondeforming Fragments for a Thin Metal Target

1. The following parameters are determined from the size, shape, and material of typical fragments, assuming a spherical shape: radius (a) and density (ρ_f).
2. The following $h/a \leq 2.2$ parameters are determined from target properties: density (ρ_t), yield stress (σ_t), and thickness (h).
3. If $h/a \leq 2.2$, the nondimensional limit velocity ($\overline{v_{50}}$) is determined from Fig. C5-4 as a function of the nondimensional target thickness (h/a).
4. Some uncertainty exists in the linear relationship shown, where the shaded area in Fig. C5-4 indicates the scatter of experimental data. For hard fragments that are less likely to deform, a lower nondimensional limit velocity should be selected. For softer fragments, a higher nondimensional limit velocity may be selected.
5. The limit velocity (v_{50}) is defined as the velocity at which a fragment will have a 50% chance of perforating a given target and is calculated as follows:

$$v_{50} = \overline{v_{50}} \cdot \dfrac{\sqrt{\sigma_t \cdot \rho_t}}{\rho_f} \quad (C5\text{-}12)$$

6. The limit velocity is compared with the fragment striking velocity (v_{st}) to ascertain target penetration.

Reinforced Concrete Target Materials for Small Metal Fragments. A crude rule-of-thumb estimate of the effectiveness of reinforced concrete walls for resisting penetration by metal fragments is to assume that 1 in. (25.4 mm) of mild steel is equivalent to 9 in. (229 mm) of reasonable-quality concrete. Spall-shielding may be used with concrete walls to reduce the effects of fragmentation.

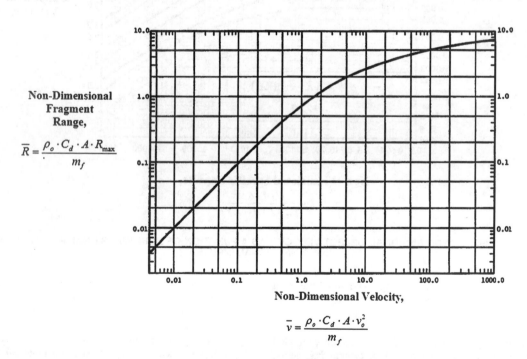

Non-Dimensional Fragment Range,

$$\bar{R} = \dfrac{\rho_o \cdot C_d \cdot A \cdot R_{\max}}{m_f}$$

Non-Dimensional Velocity,

$$\bar{v} = \dfrac{\rho_o \cdot C_d \cdot A \cdot v_o^2}{m_f}$$

FIGURE C5-3. SCALED FRAGMENT RANGE.

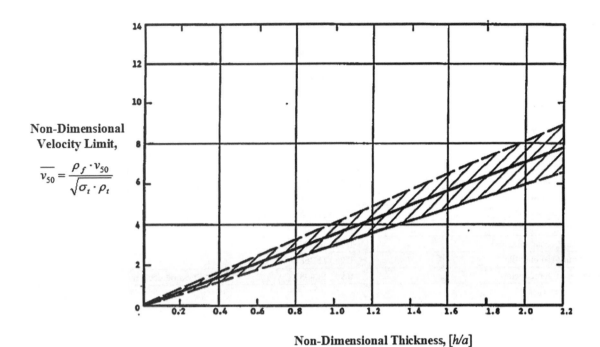

FIGURE C5-4. NONDIMENSIONAL LIMIT VELOCITY VERSUS NONDIMENSIONAL THICKNESS FOR "CHUNKY" NONDEFORMING FRAGMENTS.

1. The following properties are determined from the concrete target: compressive strength (f'_c) and thickness (h).
2. The following parameters are determined for typical fragments: weight (w_f), striking velocity (v_{st}), and diameter (d).
3. The constant C_3 is given in Table C5-2 for some typical materials.
4. The maximum penetration for steel fragments (x_f) is calculated as follows:

$$x_f = C_3 \cdot \left(3.472 \times 10^{-4}\right) \cdot \left(\frac{f'_c}{psi}\right)^{-0.5} \cdot \left(\frac{w_f}{lbf}\right)^{0.4} \cdot \left(\frac{v_{st}}{ft/s}\right)^{1.8} \cdot in.$$

(C5-13)

$$x_f = C_3 \cdot \left(3.417\right) \cdot \left(\frac{f'_c}{Pa}\right)^{-0.5} \cdot \left(\frac{w_f}{N}\right)^{0.4} \cdot \left(\frac{v_{st}}{m/s}\right)^{1.8} \cdot mm \quad \text{[SI]}$$

(C5-14)

5. Calculate the dilatational velocity of the elastic wave through concrete (C_s):

$$C_s = \left(5.16 \cdot ft/s\right) \sqrt{\frac{E_c}{psi}}$$

(C5-15)

$$C_s = \left(1.57 \cdot m/s\right) \sqrt{\frac{E_c}{Pa}} \quad \text{[SI]}$$

(C5-16)

6. The minimum thickness required to prevent spalling (T_{sp}) is calculated as follows:

$$T_{sp} = x_f \cdot \left[1 + 7.289 \cdot \left(\frac{v_{st}}{C_s}\right)^{0.333} \left(\frac{w_f/lbf}{x_f/in.}\right)^{0.457}\right]$$
$$+ \left(0.877 \cdot in.\right) \cdot \left(\frac{w_f}{lbf}\right)^{0.333}$$

(C5-17)

$$T_{sp} = x_f \cdot \left[1 + 16.161 \cdot \left(\frac{v_{st}}{C_s}\right)^{0.333} \left(\frac{w_f/N}{x_f/mm}\right)^{0.457}\right]$$
$$+ \left(13.545 \cdot mm\right) \cdot \left(\frac{w_f}{N}\right)^{0.333} \quad \text{[SI]}$$

(C5-18)

7. The minimum thickness required to prevent perforation (T_{pf}) is calculated as follows:

$$T_{pf} = x_f \cdot \left[1 + 0.325 \cdot \left(\frac{v_{st}}{C_s}\right)^{0.333} \left(\frac{w_f/lbf}{x_f/in.}\right)^{0.417}\right]$$
$$+ \left(0.877 \cdot in.\right) \cdot \left(\frac{w_f}{lbf}\right)^{0.333}$$

(C5-19)

$$T_{pf} = x_f \cdot \left[1 + 0.721 \cdot \left(\frac{v_{st}}{C_s}\right)^{0.333} \left(\frac{w_f/N}{x_f/mm}\right)^{0.417}\right]$$
$$+ \left(13.545 \cdot mm\right) \cdot \left(\frac{w_f}{N}\right)^{0.333} \quad \text{[SI]}$$

(C5-20)

8. The concrete wall thickness (h) is compared with calculated minimum required thicknesses to ascertain fragmentation protection for a concrete wall:

   If $h > T_{sp}$, then no spalling occurs.
   f $T_{sp} \geq h > T_{pf}$, then spalling occurs but no perforation.
   If $h \leq T_{pf}$, then spalling and perforation occur.

Roofing Materials. Lower limits for fragment damage to miscellaneous lightweight roofing materials based on fragment momentum are included in Table C5-5.

C5.3.2 Limits on Analytical Procedures. Inputs required for fragmentation analysis may include but are not limited to assumptions about the size, shape, and casing of the threat; the size and shape of the fragments; and location and type of objects within the surrounding site relative to the threat and building.

TABLE C5-5. FRAGMENT IMPACT DAMAGE FOR ROOFING MATERIALS

Roofing Material	Minimum Fragment Momentum (mV) for Serious Damage (lbf-sec) [N-s]	Comments
Asphalt shingles	0.159 [0.707]	Crack shingle
	1.370 [6.094]	Damage deck
Built-up roof	<0.159 [0.707]	Crack tar flood coat
	0.451 [2.001]	Crack surface of conventional built-up roof without top layer of stones
	>0.991 [4.408]	Crack surface of conventional built-up roof with 2.867 lb/ft^2 [137 Pa] top layer of stones
Miscellaneous		
1/8″ [3 mm] asbestos cement shingles	0.159 [0.707]	
1/4″ [6 mm] asbestos cement shingles	0.285 [1.268]	
1/4″ [6 mm] green slate	0.285 [1.268]	
1/4″ [6 mm] gray slate	0.159 [0.707]	
1.2″ [30 mm] cedar shingles	0.159 [0.707]	
3/4″ [19 mm] red clay tile	0.285 [1.268]	
Standing seam terne metal	0.991 [4.408]	Plywood deck cracked

Source: DOE (1992).

C5.3.3 Complex Modeling Methods. Fragmentation analysis is not well quantified and existing methods are empirical in nature. Simplified modeling methods are preferred over complex methods due to the uncertainty of input parameters required for fragmentation analyses, as described in Section C5.3.2.

REFERENCES

Baker, W. E., Cox, P. A., Westine, P. S., Kulesz, J. J., and Strehlow, R. A. (1983). *Explosion hazards and evaluation*, Fundamental Studies in Engineering 5, Elsevier Scientific Publishing Company, Amsterdam.

U.S. Department of Defense (DoD). (2008). Structures to Resist the Effects of Accidental Explosions, UFC 3-340-02, <www.wbdg.org/ccb/DOD/UFC/ufc_3_340_02.pdf> [May 12, 2011].

U.S. Department of Energy (DoE). (1992). A Manual for the Prediction of Blast and Fragment Loadings on Structures, DOE/TIC-11268 (unclassified, limited distribution), <www.osti.gov/bridge/product.biblio.jsp?osti_id=5892901> (May 12, 2011].

Chapter C6
STRUCTURAL SYSTEMS

C6.1 GENERAL PROVISIONS

C6.1.1 Purpose. The design of structures to resist blast loads typically begins with a design to resist conventional code-specified loading conditions. These structural elements are subsequently evaluated to determine their performance in response to blast loads. Where the calculated performance is found to be unacceptable, a combination of increased strength and ductile detailing may be required.

The structure's response to blast loads balance the external work performed on the structure by the external loads and the internal strain energy developed by the structural materials. The dynamic analysis accounts for the mass and inertia effects of the section to resist motion and is performed in the time-stepped analysis to predict the maximum structural element displacement that occurs during and after the blast load has been applied.

C6.1.2 Scope and Application

Seismic versus Blast. Seismic and blast-resistant design approaches share some common analytical methodologies and a performance-based design philosophy that accepts varying levels of damage in response to varying levels of dynamic excitation. Both design approaches recognize that it is cost-prohibitive to provide comprehensive protection against all conceivable events, and an appropriate level of protection that lessens the risk of mass casualties can be provided at a reasonable cost. Both seismic design and blast-resistant design approaches benefit from a risk assessment that evaluates the function, criticality, occupancy, site conditions, and design features. While there is predictability to natural hazards, this is not the case with man-made hazards. Also, the explosive threats of the future are very likely to be very different from the explosive threats of the past. Another fundamental difference between seismic and blast events are the acceptable design limits. Since earthquakes are more predictable and affect more structures than are affected by blast events, owners may be willing to accept different levels of risk relative to these different events, and this may translate into differences in acceptable design limits, as defined by allowable deformation, allowable ductility, etc.

Both seismic design and blast-resistant design approaches consider the time-varying nature of the loading function. Due to the varying nature of the seismic excitation from location to location—in intensity, duration, and frequency content of any given event—the prevailing design approach is to use an envelope spectrum that characterizes the response acceleration intensity of the fundamental modes of response. Due to the global nature of building response to earthquake motions, which generally apply base motions uniformly over the foundations, the spectral approach allows the engineer to account for site-specific features and energy dissipation mechanisms without having to identify a particular ground motion history. These seismic motions induce forces that are proportional to the mass. Blast loading is not uniformly applied to all portions of the building. Parts of the structure and components closest to and facing the point of detonation will experience the higher loading than components at a greater distance and/or not facing the point of detonation. The structure's mass contributes to its inertial resistance. As a result, the seismic loading analogies, including the concept of blast-induced base shears, must be applied with great care lest they be misconstrued to provide a false sense of protection.

Blast loads from an exterior detonation will deform the exterior bays of the structure to a much greater extent than the interior bays and, in the time frame of the applied loading, the story shears may not necessarily be distributed through the diaphragms in proportion to the framing stiffness. As a result, the ductility demands of components closest to and facing the point of detonation far exceed the demands elsewhere throughout the remainder of the structure. Although a first-mode response of the structure is used to estimate the blast-induced story shears and these story shears are often used to evaluate the lateral-load-resisting system's ability to transfer the blast forces to the foundations, these approximations may be misleading.

It should be noted that strong ground motions can also lead to progressive collapse, in which a failure of a local element in a structure leads ultimately to the failure of a disproportionate region of the structure. During the Mexico City earthquake (1985), the lack of lateral-load-resisting systems in apartment houses caused a complete collapse in the lower floors, which in turn caused the complete collapse of higher floors until entire buildings "pancaked" upon themselves.

Both seismic and blast design considerations make posttensioned, two-way flat slab structures the least desirable cast-in-place concrete framing system. Punching shear failures at the slab joint must be avoided under lateral sway deformations and these systems possess limited lateral deformation capability. However, for seismic systems, a two-way flat slab system might perform satisfactorily when it is used in combination with a stiffer shear wall system that controls the ultimate lateral sway of the structure and prevents destabilizing P-delta effects. The punching shear failure vulnerability in response to blast loading is not improved with the addition of shear walls.

Seismic excitations engage the entire lateral-load-resisting system, while resistance to blast loading is more concentrated to the structure in the vicinity of the explosion. Moment frames may provide high ductility and energy absorption when the plastic hinging of beams is spread throughout the structure. For plastic hinging to be spread throughout the structure, the beam-column connections must develop the flexural hinge in the beam and the elements must be proportioned to avoid plastic story mechanisms that would limit the inelastic energy absorption to isolated regions of the frame. This is achieved through the "strong-column weak-beam" (SCWB) approaches for both seismic and blast design. Due to the localized nature of blast events, the plastic hinging is more localized with more severe rotations than those that result from earthquake events. Concentrically braced frames (CBFs) are commonly used to resist lateral seismic and lateral wind loads on buildings. The lateral loads are efficiently resisted through axial forces. Another bracing system, the eccentric braced frame (EBF), offers lateral resistance through a combination of axial forces in the brace and shear forces and bending moments in the link beams. Both systems can be highly efficient in response to the seismic hazard. Due to the localized nature of the blast forces, the effectiveness of either CBF or EBF must be considered on a case-by-case basis.

Building configuration characteristics, such as size, shape, and location of structural elements, are important issues for both seismic and blast-resistant design. The manner in which forces are distributed throughout the building is strongly affected by its configuration. While seismic forces are proportional to the mass of the building and increase the demand, inertial resistance plays a significant role in the design of structures to reduce the response to blast loading. Structures that are designed to resist seismic forces benefit from low height-to-base ratios, balanced resistance, symmetrical plans, uniform sections and elevations, the placement of shear walls and lateral bracing to maximize torsional resistance, short spans, direct load paths, and uniform floor heights. While blast-resistant structures share many of these same attributes, the reasons for doing so may differ. For example, seismic excitations may induce torsional response modes in structures with reentrant corners, whereas these conditions provide pockets where blast pressures may reflect off of adjacent walls and amplify the blast effects. Similarly, first-floor arcades that produce overhangs or reentrant corners create localized concentrations of blast pressure and expose areas of the floor slab to uplift blast pressures. In seismic design, adjacent structures may suffer from the effects of pounding in which the two buildings may hit one another as they respond to the base motions. Adjacent structures in dense urban environments may be vulnerable to amplification of blast effects due to the multiple reflections of blast waves as they propagate from the source of the detonation. While the geology of the site has a significant influence on the seismic motions that load the structure, the surrounding geology of the site will influence the size of the blast crater and propagation of ground shock.

On an element level, the ductility demands for both seismically loaded structures and blast-loaded structures require attention to details. Concrete columns require lateral reinforcement to provide confinement to the core and prevent premature buckling of the rebar. Closely spaced ties and spiral reinforcement are particularly effective in increasing the ductility of a concrete compression element. Carbon fiber wraps and steel jacket retrofits provide comparable confinement to existing structures. Steel column splices must be located away from regions of plastic hinging or must be detailed to develop the full moment capacity of the section. Local flange buckling must be avoided through the use of closely spaced stiffeners or, in the case of blast-resistant design, the concrete encasement of the steel section. Reinforced concrete beam sections require equal resistance to positive and negative bending moments. In addition to the effects of load reversals and rebound, doubly reinforced sections possess greater ductility than singly reinforced counterparts. Steel beams may be constructed composite with the concrete deck in order to increase the ultimate capacity of the section; however, this increase is not equally effective for both positive and negative moments. While the composite slab may brace the top flange of the steel section, the bottom flange is vulnerable to buckling. Tube sections and concrete encasement are particularly effective in preventing flange buckling under load reversals.

New versus Existing Construction. New construction provides greater opportunity for blast resistance than existing construction. Structural and nonstructural elements may take advantage of debris-mitigating materials and robust construction details. Furthermore, if conceptualized during the preliminary phase of design, blast-mitigating features may be incorporated into the architectural plans, elevations, and structural framing systems. The protective design of new buildings therefore permits the selection of preferred framing systems that accommodate large, localized inelastic deformations and reversals without developing brittle failures or instability. Depending on the type of existing construction, a structure may possess sufficient inertial resistance and strength to withstand an explosive loading; however, the opportunities to retrofit the structure are limited. Lightweight FRP composites represent one type of retrofit that may be applied to the surface of existing construction to supplement tensile reinforcement and/or enhance element ductility. Due to the weight and space efficiency of FRP composite construction, structural element upgrades utilizing FRP composites are extremely effective in the blast-hardening retrofit of existing buildings where added weight and space are major constraints. Connection details can also be upgraded to accept larger forces. However, for new building construction, blast resistance may be directly incorporated more efficiently into the design of key structural elements.

C6.2 STRUCTURAL MODELING AND ANALYSIS

C6.2.1 Analytical Methods. The protective design of structures to resist explosive loads may be accomplished using a wide variety of analytical approaches. These approaches include empirical explosive test data that can be expressed as prescriptive requirements, P-I charts, single element response analysis, structural system response analysis, and detailed finite element analyses. Other numerical analysis methods are acceptable as long as they are validated for the similar structures subjected to similar loading environments. All approaches attempt to represent the behavior of the structural system in response to a short-duration but high-intensity dynamic loading. In all cases, the analytical methods must be numerically robust in order to characterize the transient blast loading pulse and the dominant frequencies of response associated with the postulated failure mechanisms. Care must be taken so as not to neglect the critical failure mechanisms that may precipitate collapse or hazardous debris. Since it is important that the selected analytical approach be capable of representing the structure's likely failure mechanism, the designer must be able to anticipate the response before selecting an analytical approach. A detailed discussion of the principles of structural dynamics may be found in UFC 3-340-02 (DoD 2008).

C6.2.1.1 Pressure-Impulse Charts. Pressure-impulse (P-I) diagrams summarize the performance characteristics of a type of structural or nonstructural element in response to a range of explosive loading (Fig. C6-1). The explosive loading parameters are expressed in terms of peak pressure and impulse and the corresponding P-I curves indicate the threshold of the different levels of protection or hazard. Typically, one end of the concave P-I curves transitions to a pressure asymptote, while the other end of the curve transitions to an impulse asymptote. These two asymptotes represent the minimum impulse for all greater peak pressures or minimum peak pressure for all greater impulses that correspond to a specified performance threshold. Typically, all combinations of peak pressure and impulse that lie below or to the left of a P-I curve correspond to a better performance condition, while all combinations of peak pressure and impulse that

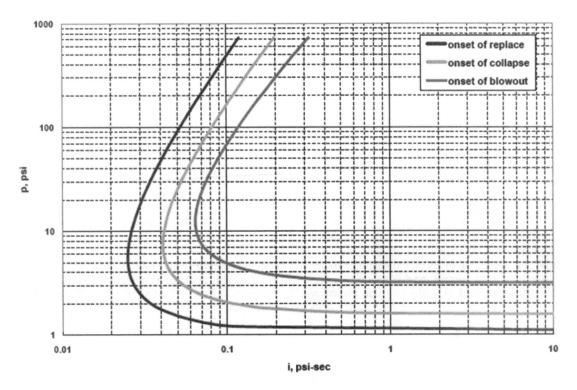

FIGURE C6-1. REPRESENTATIVE PRESSURE-IMPULSE CURVE.

lie above or to the right of a P-I curve correspond to a worse performance condition.

The P-I curves may either be derived from an analytical database of performance or may be fit from available test data. Often, the parameters are presented in nondimensional form where the peak pressure and impulse axes are normalized so that a given set of curves and plots may be used to represent the performance of a greater range of element responses to a wide variety of explosive threat conditions. P-I curves may only be used if they are generated using the general principles described in this Standard or by approved testing in accordance with any of the following: (1) government test data acquisition procedures; (2) test data collection following a national standard; or (3) test data collection in accordance with Chapter 10 of this Standard. In the following example of a P-I curve, all combinations of pressure and impulse that are to the right of a given curve are associated with the specified level of performance.

P-I curves may be used for simplified evaluations of secondary elements but should not be used for the design of primary structural elements, except for one-story reinforced masonry bearing wall structures in response to external blast loads (Section 6.3.1.6), when the P-I curve is developed according to the provisions of this Standard or when otherwise acceptable to the authority having jurisdiction.

C6.2.1.2 Single Element Response Analysis. Individual elements may be analyzed independently by means of either single-degree-of-freedom (SDOF) or multi-degree-of-freedom (MDOF) inelastic dynamic methods. These analytical methods require the use of explicit time-step numerical integration algorithms, ranging from constant velocity to linear acceleration techniques, along with the appropriate constitutive relations that represent both the elastic and inelastic behavior of the element. SDOF models are inherently simple and the accuracy of the response calculation depends on the approximations that are used to characterize the dynamic response of the element. MDOF models may be significantly more complex and may be developed using finite element methods (FEM) to derive the mass and stiffness matrices.

SDOF methods are adequately described in a variety of structural dynamics textbooks and U.S. Army technical manuals. The accuracy of this approach depends on selecting the appropriate SDOF system to represent the governing failure mechanism of the element. While this approach is the most common method applied to the design and analysis of elements to blast loading, it requires considerable experience to make sure a critical failure mechanism is not overlooked. Typically, flexural modes of response are represented by the SDOF models; however, direct shear and instability may prove to be more critical response characteristics. The dynamic loads that may be applied to SDOF analyses of individual structural elements may be based on tributary areas or may be derived from the dynamic reactions of the secondary elements that frame into a primary element. The application of dynamic reactions from subsidiary elements accounts for the frequency of response and limiting capacity of the subsidiary elements, but must also include the appropriate amount of additional mass. The dynamic reaction forces from a series of subsidiary element analyses may be converted into an equivalent dynamic load by equating the work performed by the individual reactions through a representative deformation.

MDOF analyses may be performed using a variety of dynamic response analysis software programs; however, all analyses must be performed at a short time-step to guarantee numerical stability, to resolve the frequency content of the model, and to accurately represent all potential modes of failure. Nonlinear dynamic finite element analyses provide the most comprehensive means of determining the self-consistent stiffness and inertial properties, the limiting plastic limits, and the governing failure mechanisms. Furthermore, nonlinear geometric algorithms provide the means to include the destabilizing P-delta effects due to the coupling of the large lateral transient deformations and the static gravity loads. MDOF models accurately represent the spatial distribution of dynamic pressures and dynamic reactions from subsidiary element analyses without having to generate an equivalent dynamic load.

Damping effects will be minimal when significant amounts of strain energy may be dissipated through inelastic deformation. Damping should be ignored for these conditions. When the response is nearly elastic, levels of damping typically considered for seismic response may be used.

C6.2.1.3 Structural System Multi-Degree-of-Freedom Finite Element Response Analysis. SDOF responses and MDOF models of individual elements do not account for the interaction between interconnected elements, the phasing of their responses, and the flexibility of the actual boundary conditions. These simplified approaches rely heavily on engineering judgment for the selection of additional mass. As a result, the idealizations represented by SDOF and MDOF analyses of individual structural elements do not accurately account for the dissipation of energy as the entire structure is deformed and suffers damage. Structural system response calculations do consider the relative flexibility and strength of the interconnected structural elements and provide a more accurate distribution of blast load and element response. Most importantly, however, structural system analyses consider the phasing of the responses between the different structural elements and provide the most accurate representation of boundary conditions and inertial resistance.

C6.2.1.4 Explicit Linear or Nonlinear Finite Element Analysis. Explicit dynamic finite element methods are demonstrated to be effective to determine the performance of blast-resistant structures; however, the use of these methods requires skilled and experienced modelers. Explicit dynamic finite element methods capture the high-frequency characteristics of the shock wave loading and structural response. Explicit formulations of the equations of motion express the displacement at a given time-step t_{j+1} in terms of displacements, velocities, or accelerations at previous time-steps. This approach captures the high-frequency effects of the shock loading and easily allows for both material and geometric nonlinear effects; however, all explicit formulations have a critical time-step, above which the solution becomes numerically unstable. The critical time-step is related to the wave speed of the material and the least dimension of the finite element model. The equation of motion is therefore solved at each node using the current geometry and material properties at that location and point in time. An implicit formulation of the equations of motion expresses displacement of each node at a given time-step t_{j+1} in terms of all the displacements, velocities, or accelerations at that time-step t_{j+1}. The equations of motion for the entire system must be solved simultaneously; this is usually done using matrix methods and is a very efficient way of calculating the response of linear elastic systems. For nonlinear systems, however, an explicit formula (predictor) is usually used to estimate the response at the end of each time-step, and this is followed by one (or more) correction to improve the results. Such an approach, while necessary to obtain a more correct analysis, can be computationally demanding and inefficient. Furthermore, even for elastic analysis, if the shock or high-frequency response of the system is needed, the time-step must be made sufficiently small to capture such effects while avoiding undue numerical damping. Implicit schemes are least efficient in such cases.

C6.2.2 Materials. Material properties, in response to blast loading, are affected by extremely high loading rates, high confining pressures, and large inelastic deformations. More sophisticated finite element representations of materials capable of explicit analysis are available but these complex methods are generally not required for the design of new construction. Explicit, nonlinear dynamic analysis methods can be sensitive to changes in constitutive material property assumptions and small changes in support boundary conditions, and the accuracy of the results is dependent on the modeling techniques used.

C6.2.2.1 Material Strength Increase. Materials such as concrete and steel are typically stronger than the minimum values specified in the construction documents. ASTM specifications and codes define the minimum properties for materials, such as in A706-06a (ASTM 2006a) reinforcing bars and A992 (ASTM 2006b) structural steel where the maximum yield strengths are also defined. The actual materials being installed have strengths that exceed these minimum requirements. For the analysis and design of reinforced concrete and structural steel elements for flexure, it is generally acceptable to increase the steel materials minimum yield strength by 10%. Ignoring a material's average strength and dynamic strength increase generally results in increased factors of safety for bending elements; however, underestimating the effects of material strength increase factors is unconservative for shear and connection design.

Structural elements subject to dynamic blast loads exhibit higher strengths than do structural elements subject to static loads. Testing has shown that materials experience an increase in strength when subject to high-strain-rate dynamic loadings, such as during an explosion. While underestimating dynamic strength increase effects results in increased factors of safety for flexural element design, it is unconservative to underestimate the dynamic increase effects when calculating element resistances to determine design loads for shear and connection design.

When combining material average strength effects with dynamic strength increase effects, the material's yield strength used for determining the resistance of an element should not exceed the ultimate dynamic strength of the material for flexural design.

Allowable dynamic increase factors for element connection designs are generally set lower than for flexural elements due to the likelihood that the rise time of loading in a connection is significantly longer than for the flexural element and for increased conservatism in the connection design.

C6.2.3 Modeling of Elements. The performance of structural elements in response to blast loading should be based on the results of dynamic inelastic analyses. A wide range of dynamic analyses may be performed, ranging from SDOF calculations to explicit dynamic finite element calculations. Explicit finite element methods capture the high-frequency shock loading and the high-frequency response of nonlinear systems. Regardless of the level of analytical rigor, these calculations should include the appropriate failure mechanisms, boundary conditions, and all concurrent states of loading. The significance of the different failure mechanisms and response characteristics will depend on the characteristics of the blast loading, the type of structural element, the structural details, and performance of the element materials.

C6.2.3.1 Flexure. Although inelastic behavior is not generally permitted under service loading design for structural elements, inelastic behavior is appropriate for the economical design of structures subject to a blast loading. The effects of rebound are not common in the conventional static design of structures and must be addressed in the design and detailing of the sections. At a minimum, the dynamic response analyses must be carried out to a sufficient duration of response in order for the full effects of rebound to be determined.

Section 6.2.3.1 requires flexural elements to develop their flexural yield capacities and therefore prevents brittle modes of failure from limiting the flexural capacity of the element. Elements that require high levels of protection may remain elastic in response to their design threats; nevertheless, their detailing should not prevent the sections from developing larger deformation in response to more intense loading. Therefore, Section 6.2.3.1 prevents all flexural elements and their connections from developing brittle failure modes that might prevent the section from developing its plastic capacity. For the special case where elements are designed to resist blast loads elastically, a factor of safety is specified for the design of the nonductile modes.

C6.2.3.1.1 Flexure in Structural Steel Elements.
Bracing and flange width-to-thickness ratios have a considerable effect on the ultimate resistance capacities of steel elements. If adequate bracing is not provided, then the element may buckle prior to developing the full plastic resistance determined by analysis, resulting in a nonductile response with less energy absorption. When width-to-thickness ratios are used to establish the section as "compact" for analysis, then the width-to-thickness ratios should take into account the amount of anticipated rotation ductility demand and should be based on the expected material strength. Although the conventional definitions for compact sections are sufficient for levels of protection I and II, the definitions for seismically compact sections are required for levels of protection III and IV.

A factor in ANSI/AISC 341-05 (AISC 2005) that defines the ratio of expected and specified yield strengths may be combined with the dynamic increase factors for blast-resistant design.

Steel elements generally exhibit large rebound responses and should be adequately braced. However, finite element studies of steel sections in response to short-duration uplift blast pressures that include the beneficial effects of rebound and gravitational restoring forces do not demonstrate the susceptibility to local flange buckling, and advanced analytical studies may be used to demonstrate the bracing requirements.

C6.2.3.1.2 Flexure in Concrete Elements.
Depending upon the magnitudes of the blast loading, permissible deformations, and ductility requirements, one of following three types of reinforced concrete cross sections can be utilized in the analysis of blast-resistant concrete flexural elements in the plastic range (Fig. C6-2). Adequate confinement of the concrete core is critical to the performance of the Type II and III elements.

Elements where a ductile response is not required:

Type I: The concrete compression block is effective in resisting moment. The concrete cover over the reinforcement of the element remains intact. The element has reinforcing in each face.

Elements required to have a ductile response:

Type I: The concrete compression block is effective in resisting moment. The concrete cover over the reinforcement on both surfaces of the element remains intact. The element has reinforcing in each face and stirrups if required by design for far-range response, and has stirrups for close-in-range response.

Type II: The concrete compression block is crushed and not effective in resisting moment. Compression reinforcement equal to the tension reinforcement is required to resist moment. The concrete cover over the reinforcement on both surfaces of the element remains intact. Stirrups are required for far-range response and close-in-range response.

Type III: Elements in systems capable of tension membrane action; the concrete compression block is crushed and the concrete cover over the reinforcement on both surfaces of the element may be completely disengaged. Compression reinforcement equal to the tension reinforcement is required to resist moment. Reinforcement is tied together with stirrups.

C6.2.3.1.3 Flexure in Masonry Elements.
In lieu of developing the required strength to resist blast loads, existing unreinforced non-load-bearing masonry partition walls may be protected with the application of an elastopolymer (or other catch system) that retains debris. The effectiveness of the elastopolymer shall be determined through applicable explosive test data or advanced finite element analysis.

C6.2.3.2 Shear, Axial, and Reaction Forces in Bending Elements.
Ductile design and detailing provides improved performance in response to extreme load conditions, which may not be reliably represented by the design basis threat. The 1.5 factor on brittle modes is intended to force a ductile failure.

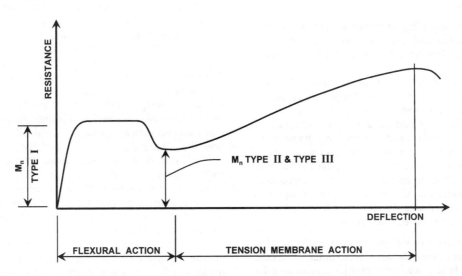

FIGURE C6-2. REPRESENTATIVE MOMENT-DEFLECTION RELATION FOR TYPE I, II, AND III ELEMENTS.

C6.2.3.2.2 Shear in Reinforced Concrete Elements. See UFC 3-340-02 (DoD 2008) for discussion of brittle modes of failure and the effects of close-in explosions.

C6.2.3.3.1 Compression Elements. If a dynamic analysis is performed to determine the axial forces in the ductile flexural elements with compression due to blast loading, which considers the dynamic nonlinear response with respect to arrival time, ductile flexural elements with compression can be designed for flexure and axial forces using the results of the dynamic analysis in lieu of the sum of the resistance reactions from the members framing into the columns.

C6.2.3.4 Instability. Instability of the overall structure or premature buckling of individual elements limits the ductility of the structure. Instability of the structure occurs when loads and deflections of the structure or structural elements become excessive due to blast loading. Instability can also occur when members and secondary bracing elements, such as floor diaphragms, are destroyed by blast effects. In blast-resistant construction, exterior façade typically spans from floor to floor so they do not add tributary loads to the exterior building columns; however, the blast load reactions from the exterior façade must be applied to the supporting structure and the stability of the entire structure must be checked. This is particularly important for light-weight structures.

C6.2.3.4.1 Element and System Stability. P-delta effects must be considered for all structural elements that sustain the combined effects of lateral deformations and axial loading. Axial loads must be considered in the analysis of columns that sustain large lateral deformations. Similarly, the buckling of bracing beams must be considered if the floor diaphragms are severely damaged.

C6.2.3.4.2 Local and Lateral-Torsional Buckling. Analytical studies are likely to demonstrate that rebounding elements and gravitational restoring systems are not susceptible to local buckling.

C6.2.3.4.4 Progressive Collapse. The major purpose of this Standard is to minimize the loss of lives by making structures insensitive to local damage resulting from unforeseen extraordinary events. This is achieved by stating performance objectives or specifying prescriptive design requirements. The design to prevent progressive collapse is addressed by this Standard for three primary reasons:

1. Owners might be unwilling or unable to provide for fully "hardened" facilities and, therefore, might accept risk of structural damage but are unwilling to accept the risk of collapse.
2. Threats that change with time may not be adequately anticipated and owners might want a level of protection against collapse in the event that the threat is underestimated.
3. Some of the technology associated with blast phenomena, building response to blast, and progressive collapse resistance simply is not developed to the level necessary to provide the same level of confidence in designs as for other loading forms. Although advanced analytical capabilities and design guidance do exist, there is a greater degree of uncertainty associated with the design of hardened structures as compared to conventional design requirements.

It is therefore important to evaluate the potential for blast damage, resulting from a design-level event, to initiate a broader extent of collapse. It is also important to offer a procedure for specific local resistance that provides a hardened structure that will resist the effects of the explosive event, which is different from the threat-independent progressive collapse approaches that may be independently specified in a building code. The Standard therefore enables the project team to estimate the damage zone so that the owner can appreciate the implications of his/her decisions and the designer can provide viable means to arrest collapse. Although the vertical progression of collapse for buildings shorter than three stories is excluded from the UFC 4-023-03 (DoD 2010), the lateral progression of collapse for short buildings may still be a concern.

As a minimum, the Standard encourages structural ductility and redundancy as essential elements of an effective blast-resistant design—regardless of whether progressive collapse is addressed directly. This pertains to both global building ductility demands (involving all elements of the framing system) and local ductility around the impacted frames of the building. In this spirit, the Standard references some approaches for detailing and minimum tie forces throughout the portions of the structure that are expected to participate during an event. This is threat-independent and therefore has little direct relevance to a particular damage pattern, but is also simply good practice. It also addresses a practical limitation to a designer's ability to estimate the extent of damage much beyond the outer column line in response to an exterior explosion. Accurate analysis of damage patterns requires advanced analysis to calculate the infill pressures and advanced dynamic inelastic analyses to determine the structural response. These methods are often beyond the scope of work for the vast majority of projects, and even the simplified procedures require a significant level of expertise. The evaluation of uplift resulting from infill pressures must be considered when the exterior façade is overwhelmed by blast loading.

Two of the three approaches in the literature are threat-independent (prescriptive detailing and alternate path) and only one (specific local resistance) addresses the threat-induced state of damage. The Standard acknowledges the threat-independent approaches for structures where no explosive threat is specified or where a nominal threat is specified to provide a nominal umbrella of protection, as they quantify a level of continuity and detailing that produces a more robust structure. This is the approach taken by the General Services Administration (GSA) and the U.S. Department of Defense (DoD). However, these threat-independent methods are most appropriate when the damage associated with the explosive threat will be localized.

Studies are currently underway to demonstrate the relationship between blast damage and progressive collapse. Calculations are being performed to demonstrate the effectiveness of the threat-independent approaches in response to the damage resulting from explosive detonations; although the threat-independent approaches are most effective when the extent of initiating damage is relatively localized, results show they do not provide an effective umbrella of protection in response to the much-larger-than-design-level-threat that might occur. Therefore, the initiating condition should also include removal of certain adjacent elements (beams, slabs, etc.) in this case, since clean separation of a column from beams is unlikely.

For blast-resistant design, a threat must be identified in order for the initial state of damage to be estimated. The residual capacity of the structure can be evaluated and the effectiveness of the system to prevent the progression of collapse may be determined. This approach is similar to that of performance-based seismic design codes for which the engineer selects the hazard (event) and the desired performance, such as low levels of damage (immediate occupancy) in response to a small earthquake and life safety protection (damage short of complete collapse) in response to an extreme event. Either the structural elements must be designed to resist the specified threat and maintain load-carrying capacity, or a permissible level of localized damage may be permitted as long as the damaged structure is still capable of preventing a progression of collapse. This corresponds to the "extreme event" in response to which the structure must remain standing for life safety and evacuation of the occupants. Engineers should be provided with options to address the issue and be allowed to pursue any rational option so long as analyses demonstrate the effectiveness of the approach.

A prescriptive method may be applied to cases where no threat is specified. An "arbitrary" (but well-conceived) damage volume may be prescribed as a function of threat, performance expectation, and/or level of specific local resistance provided in the design, perhaps as measured by some sort of "demand/capacity" ratio—for example, how much excess (or lacking) blast resistance is intentionally provided. Similar to the DoD criteria, the designer would then be required to demonstrate that the structure is stable with the specified damage. This approach could be implemented without conducting multiple blast assessments, yet provide a measure of protection that is calibrated to the effects of "Low," "Medium," and "High" incidents. While this approach may be applicable to the majority of designs, there will be circumstances when owners and designers will need to pursue analyses that will more accurately determine damage zones and collapse potential. There needs to be clear guidance to indicate when these circumstances need to be addressed. For these cases, the Standard may guide the design team to select a design threat in order to determine the damage state. The damage volume may be related to "Low," "Medium," and "High" incidents, and the designer may be required to bridge over such damaged volumes by demonstrating that the structures can develop the required tie forces. Full nonlinear analysis, SDOF, and P-I methods can be used to define the initial state of damage and estimate the damaged volume.

Finally, when evaluating the structure to determine its response to either the removal of a column or the damage state following a detonation, the Standard encourages the use of the most accurate analytical methods that are consistent with the initial damage volume based on the size of the explosive event. Only inelastic, large-displacement, dynamic analyses may simulate the actual structural behavior. Material nonlinearity may be sufficient for cases where deformations remain small and geometric nonlinearity is not significant; however, this cannot be predicted without a fully nonlinear dynamic analysis. Simplified equivalent static methods are not demonstrated to be consistently conservative and their use is discouraged by the Standard. However, since most projects are unlikely to support the use of dynamic inelastic analytical methods—clients rarely value the benefit and are rarely willing to pay for high-end computations—simplified linear static approaches must be permitted with appropriate safeguards. The following summarizes the objectives of the Standard:

1. Specify prescriptive means to accomplish what are considered the minimum requirements for ductility and redundancy.
2. Specify minimum design-basis damage zones that are tied to basic threat size, performance expectations, and the level of hardening intentionally designed into the system.
3. Until the state of the practice advances, allow progressive collapse designs using commonly accepted approaches (linear) when the initial conditions match the assumptions that form the bases of those approaches with appropriate safeguards.
4. Require more accurate progressive collapse analysis approaches (various forms of nonlinear dynamic) when

initial conditions do not match assumptions consistent with simplified approaches.
5. Allow designers to pursue more rigorous approaches in any circumstance when conditions warrant or the owner so requests.

C6.2.4 Connections and Joints. Connections shall be designed, detailed, and constructed such that the governing failure mode of the connection is ductile. Fracture of bolts and welds shall not be the governing failure mode of connections. In order to guarantee that the structure will develop ductile modes of failure, components of connection details that may be vulnerable to brittle modes of failure must be designed to resist forces that are significantly greater than the yield capacity of the elements. The reaction forces should therefore be amplified by a sufficient factor to preclude a brittle connection failure. The 1.5 factor on brittle modes is intended to force a ductile failure.

C6.2.5.1 Empirical Data, Ray Tracing, Computational Fluid Dynamics. The most common means of calculating blast loads that are produced by a specified explosive threat is by means of empirical relations. These relations, often referred to as Kingery-Bulmash equations, may be represented in a variety of forms. They are often found in the literature in the form of nondimensional log-log plots or within special-purpose software, such as UFC 3-340-02 (DoD 2008). These empirical relations provide peak pressures and the corresponding impulses for varying angles of incidence between the orientation of the blast waves and the loaded surface. Reflections that may result from multiple surfaces, such as floor and ceiling slabs relative to walls, may be determined using a ray-tracing approach. This approach accumulates the effects of multiple reflecting waves and constructs the "saw tooth" pressure time histories, from which the corresponding impulse may be determined by numerical integration. However, the results of these calculations are extremely sensitive to the number of rays selected for consideration, and comparisons with empirical relations must be made in order to verify the accuracy of the results.

Computational fluid dynamics (CFD) methods represent the state of the art in calculating the blast wave loading on structural elements. These methods discretize the space encompassing the explosive threat and the loaded surface, and propagate the waves through the conservation of energy and momentum. CFD approaches require appropriate numerical algorithms in order to represent the equations of state that define the source of the explosive event and resolve the high-frequency shock waves as they propagate through the various media. CFD methods are often required in order to calculate the extreme pressures resulting from near-contact explosive threats and for complex geometries in which multiple reflections may affect the calculated results. Coupled calculations represent the state of the art for interactively determining the blast pressures on loaded surfaces and the resulting structural response. Coupled calculations minimize the errors associated with idealizations of loaded structures as either rigid bodies or dense matter with no defined structural integrity.

C6.2.5.4 Negative Phase. Analysis for negative phase blast loading can generally be neglected in concrete elements that are reinforced in each face with sufficient area of reinforcing to resist the rebound forces.

Analysis for negative phase blast loading can generally be neglected in steel elements that have sufficient flange bracing that a hinge can form when the element reverses direction and rebounds.

When concrete elements are not reinforced in each face for rebound forces or when steel elements are not braced such that a hinge can form when rebounding, and when the negative phase blast pressure duration exceeds one-half of the period of the inelastic response of the element, then elements should be analyzed for rebound forces in conjunction with the negative phase blast pressures. Other construction materials should be similarly detailed and analyzed for rebound and negative phase blast pressures.

C6.2.5.5 Element-to-Element Load Transfer. Energy methods may be used to equate the amount of work performed by a concentrated load through its deformation and an equivalent uniform load through a corresponding deformation.

C6.2.5.6 Tributary Loads. Tributary loads are most accurately represented by MDOF systems in which the phasing of the load and the mass of the framing elements are appropriately represented.

C6.2.5.8 Clearing Effects. Blast pressures that are applied to the surfaces of buildings or building components will be diminished to the incident pressures in the presence of an edge or opening. The clearing time at which this decay occurs is a function of the blast wave speed and the clearing distance to the edge or opening.

C6.2.6 Mass. The use of SDOF methods are documented in textbooks on structural dynamics and in UFC 3-340-02 (DoD 2008).

C6.3 STRUCTURAL DESIGN

C6.3.1 Structural Systems. Analysis of a structural system is dependent on the structural system. It is common for codes and design guidance to organize the seismic treatment of structures according to the structural systems as bearing wall systems, building frame systems, moment-resisting frame systems, and dual systems. For appropriate treatment of blast analysis, the following systems are selected in this Standard: steel moment frame, steel braced frame, concrete moment frame, concrete frame with concrete shear walls, precast tilt-up with concrete shear walls, and reinforced masonry bearing walls with rigid diaphragms. The systems are described in detail as they pertain to blast analysis. Primary structural systems that are associated with a large influence area require a greater level of protection than secondary structural systems that may result in more localized consequences. Greater attention should be paid to both the analysis and detailing of primary elements.

C6.3.1.1 Steel Moment Frame Systems. The steel moment frame (SMF) is a moment-resisting frame system. All vertical and lateral loads are carried by columns. Lateral forces are resisted by flexure of the frame columns and beams by way of rigid or semi-rigid connections between the frame beams and columns. SMFs support both gravity and lateral loads. SMFs are sometimes located on the perimeter of the structure.

The SMF is ductile and redundant if numerous bays of moment-resisting frames are provided. The stiffness of the SMF for lateral loads is generally low due to the reliance on flexure of the SMF members to resist lateral forces.

A SMF system is a relatively light-weight system with low inertia and little resistance to uplift.

Although the shear connection must be stronger than the load that can be delivered in flexure for plastic behavior, *this is only required* up to the blast load requirements when the members remain elastic.

C6.3.1.2 Steel Braced Frame Systems. The steel braced frame (SBF) is a building frame system. All vertical and some lateral

loads are carried by columns. The SBF consists of vertical members (columns) and vertically diagonal members between the columns. While the columns in the SBF may support both gravity and lateral loads, the diagonal members generally provide resistance for lateral loads only. The number of SBFs (or group of SBFs) in a structure is relatively low and the location of SBFs is generally widely separated. SBFs are often located on the interior of the structure to avoid conflict with the cladding system and to make use of large gravity loads to offset lateral overturning loads.

The SBF is generally considered to provide a medium level of ductility and is not considered highly redundant. Ductility, however, can be increased through the use of eccentric braced framing. Large diaphragm forces can develop with an SBF system due to the low number of SBFs. The stiffness of the SBF is high due to the reliance on axial loading of the bracing members to resist lateral forces.

An SMF system is a relatively light-weight system with low inertia and little resistance to uplift.

Although the shear connection must be stronger than the load that can be delivered in flexure for plastic behavior, *this is only required* up to the blast load requirements when the members remain elastic.

C6.3.1.3 Concrete Moment Frame Systems. A concrete moment frame (CMF) system is a moment-resisting frame system. All vertical and lateral loads are carried by columns. Lateral forces are resisted by flexure of the frame columns and beams by way of rigid connections between the frame beams and columns. CMFs can support both gravity and lateral loads. The CMFs are often located on the perimeter of the structure.

The CMF is ductile and redundant if numerous bays of moment-resisting frame are provided. Ductility can be improved by utilizing special detailing of the concrete reinforcement. CMFs generally do not develop large diaphragm forces due to the extent and location of the frames. The stiffness of the CMF for lateral loads is considered moderate due to the reliance on flexure of the CMF members to resist lateral forces.

A CMF system is a relatively heavy-weight system with high inertia and substantial resistance to uplift.

C6.3.1.4 Concrete Frame with Concrete Shear Wall Systems. The concrete frame with concrete shear walls (CFSWs) is a building frame system. The CFSW system consists of columns and walls that support both gravity and lateral loads. The number of shear walls (or groups of shear walls) in a CFSW structure is generally low and the location of the shear walls is generally widely separated. Shear walls may be located on the interior of the structure to avoid conflict with the cladding system.

The CFSW is generally considered to provide a medium level of ductility and low redundancy for resistance to lateral forces. Ductility can be improved with special detailing of the concrete reinforcement. Large diaphragm forces can develop in a CFSW system due to the low number and wide spacing of the shear walls. The stiffness of the CFSW system is high due to the reliance on the shear walls to resist lateral forces.

A CFSW system is a relatively heavy-weight system with high inertia and substantial resistance to uplift.

C6.3.1.5 Precast Tilt-Up with Concrete Shear Wall Systems. Tilt-up is a construction technique of casting concrete elements in a horizontal position at the jobsite and then tilting and lifting the panels to their final position in a structure. A precast tilt-up system with concrete shear walls (TILT) is a bearing wall system with a steel-framed floor or roof system.

The TILT system consists of columns and walls that support gravity loads. Columns, if present, usually are constrained to the interior of the structure. The bearing tilt-up walls in a TILT system are also shear walls that resist lateral loads. The number of shear walls is generally extensive, often including the entire perimeter wall.

The TILT system is considered to provide a medium level of ductility and is highly redundant for lateral force resistance. The walls in a TILT system are slender, however, and are sensitive to local lateral loads and gravity loads. TILT systems are generally only one or two stories tall and rarely more than four stories, such that, when combined with gravity loads, overturning in the plane of the wall is usually not an issue. Diaphragm requirements can be significant due to the potentially large distance between shear walls. The stiffness of the TILT system is high due to the reliance on the shear resistance of the walls to resist lateral forces.

A TILT system is a relatively light-weight system with low inertia and little resistance to uplift. The connections for TILT systems should be explicitly evaluated for negative phase loading.

C6.3.1.6 Reinforced Masonry Bearing/Shear Walls. The reinforced masonry bearing wall system (MBW) often has no columns, while the walls provide gravity support and provide resistance to lateral loads. For this reason, the number of walls is generally large and the walls are located uniformly throughout the structure. The floor system in the MBW is cast-in-place concrete or precast concrete plank with topping, which may achieve a rigid diaphragm depending on the aspect ratio of the diaphragm and the diaphragm thickness.

MBW systems are considered to have moderate ductility but are highly redundant for lateral force resistance. Redundancy is low, however, for gravity loads.

The negative phase has a greater effect on the response of masonry wall systems that are subjected to very-short-duration blast loading, which generally corresponds to detonations at short standoff distances. As a result, P-I curves that would produce an impulse asymptote in the absence of the negative phase tend to overstate the capacity in this loading regime. This "layover effect" is most pronounced for low-resistance masonry walls. Pressure-impulse (P-I) curves that are developed from explosive testing that "layover" to the right in the impulsive region, due to the negative phase impulse, attribute a greater capacity of the masonry walls in response to blast loading than considering positive phase loading alone. While this may be used to evaluate the performance of existing structures, the negative phase and the associated "layover effects" should not be considered during design. Due to uncertainties in the nominal design threat scenarios, reliance on the beneficial effects of negative phase may result in an inadequate design.

C6.4 RESPONSE CHARACTERISTICS

C6.4.1 Close-In Effects. The exposed surface of a structure, or element of a structure situated immediately adjacent to the explosion, is usually subject to close-in high-pressure effects. In such situations, the initial forces acting on the structure are extremely high pressures that have been greatly amplified by their reflections on the structure. The durations of the applied loads are short, particularly where complete venting of the explosion occurs. High-pressure blast load durations are short in comparison to the time it takes the elements of the structure to reach maximum deflections from the blast loading. Therefore, structures subjected to blast effects in the high-pressure range

can, in certain cases, be analyzed for the impulse rather than the peak pressure associated with longer-duration blast pressures.

Fragments associated with the high-pressure range usually consist of high-velocity primary fragments from the explosive casing breakup or the acceleration of objects positioned close to the explosion. When people or other assets such as equipment are being protected, then the possibility of these primary fragments penetrating the structure should be completely eliminated. Also associated with the "close-in" effects of a high-pressure design range is the possible occurrence of spalling and scabbing of concrete elements and postfailure fragments. Spalling and scabbing result from the disengagement of the concrete cover over the reinforcement on the back side of a concrete structure (the side away from the explosion). Spalling can occur immediately after the blast pressures strike the concrete element and is caused by the high-intensity blast shock wave reflecting off of the back surface. Scabbing is associated with the disengagement of the concrete due to the element undergoing large displacements. Depending on their velocity and size, concrete fragments can be a hazard to people and equipment.

When designing blast-protective structures or blast-containment structures for close-in effects, the structure should not be breached. Protective and containment structures adjacent to occupied areas should be constructed with double walls separated by a void of sufficient size that allows for the deflection of the blast wall, unless it can be demonstrated that the deflection of the wall will not cause the concrete from the primary wall to spall or scab or wall-mounted equipment or wall finishes on the opposite side to disengage and become projectiles resulting in injury to the occupants.

Specific details are very often critical to the calculated loading and to the calculated performance of critical structures in response to close-in effects. Nonconservative conclusions may be drawn if the specific design condition is not represented within the database calibration range of simplified or empirical tools. Comparison of *CONWEP* and CFD calculations of close-in blast loading show that *CONWEP* underpredicts loads for scaled ranges less than 1. Furthermore, finite element response calculations of columns that are subjected to both the *CONWEP* and CFD loading functions show the simplified loading underpredicts the structural damage. Tests that are conducted at subscale overpredict the capacity of full-scale structures—full-scale structures tend to respond at lower strain rates, and size effects influence fracture; hence, full-scale structures are weaker than the subscale counterparts. Similarly, for afterburning HE, larger charges are more efficient and produce more energy per pound than smaller charge weights. Therefore, the knowledge and experience of the engineer is at least as important as the tools used, and the engineer who understands the limits of the tools may be able to compensate for the limitations with conservatism. This may not be critical where lower levels of protection are required or for conditions where a threat-independent progressive collapse design allows the structure to accept localized damage without suffering catastrophic consequences. However, an exact test of the design condition or a physics-based approach is recommended for conditions that are particularly sensitive or for which failure mechanisms may be masked unless the most detailed analyses are performed. This will be the case where the protection of high-LOP (IV) structural components in response to close-in detonations is sensitive to localized details, which determine the failure mechanisms, and where the empirical tools are not specific to these localized details. The design for close-in effects should be in accordance with UFC 3-340-02 (DoD 2008).

C6.4.2 Far-Range Effects. Structures subjected to blast pressures associated with the low-pressure range sustain peak pressures of smaller intensity than those associated with the high-pressure range. The duration of the blast load in some cases can exceed the response time of the structure. The response of structural elements analyzed for the low-pressure range depends on both the pressure and impulse of the explosion. Low-pressure, far-range explosions are often the case for building structures subject to a terrorist attack. Fragments may be formed from the breakup of the structure. These secondary fragments generally have a large mass but their velocities are generally much less than those of primary fragments.

REFERENCES

American Institute of Steel Construction (AISC). (2005). Seismic Provisions for Structural Steel Buildings, ANSI/AISC 341-05. AISC, Chicago, Ill.

ASTM International (ASTM). (2006a). Standard Specification for Low-Alloy Steel Deformed and Plain Bars for Concrete Reinforcement, A706/A706M-06a. ASTM, West Conshohocken, Pa.

ASTM. (2006b). Standard Specification for Structural Steel Shapes, A992/A992M-06a. ASTM, West Conshohocken, Pa.

Biggs, J. M. (1964). *Introduction to structural dynamics*. McGraw-Hill, New York.

Clough, R. W., and Penzien, J. (1993). *Dynamics of structures*, 2nd ed. McGraw-Hill, New York.

General Services Administration (GSA). (2003). Progressive Collapse Analysis and Design Guidelines for New Federal Office Buildings and Major Modernization Projects. GSA, Washington, D.C.

International Code Council (ICC). (2006). 2006 International Building Code. ICC, Washington, D.C.

U.S. Department of Defense (DoD). (2010). Design of Buildings to Resist Progressive Collapse, UFC 4-023-03, <www.wbdg.org/ccb/DOD/UFC/ufc_4_023_03.pdf> [May 12, 2011].

DoD. (2008). Structures to Resist the Effects of Accidental Explosions, UFC 3-340-02, <www.wbdg.org/ccb/DOD/UFC/ufc_3_340_02.pdf> [May 12, 2011].

Chapter C7
PROTECTION OF SPACES

C7.2 WALLS AND SLABS ISOLATING INTERNAL THREATS

C7.2.3 Design Provisions for Walls and Slabs Isolating Internal Threats. Frangible surfaces that facilitate venting should be used wherever protected spaces must be isolated from the effects of an internal detonation. The effectiveness of hardened partitions and hardened slabs is improved when internal detonations are permitted to vent, thereby minimizing the intensity and duration of the gas pressures. Prescreening of vehicles and parcels outside the footprint of the protected space similarly reduces the likelihood of large quantities of explosives that may evade detection.

C7.2.4 Stairwell Enclosures. Following an explosive event, the occupants of a building will attempt to evacuate the building and should be provided with a safe means of egress. This section provides design provisions for stairwell enclosures in buildings with controlled access. Even if controlled access does not exist, these design recommendations will increase protection to building occupants.

C7.3 SAFE HAVENS

C7.3.1 Design Considerations. To perform an assessment of an existing structure or a new structure to be used as a safe haven, the building owner or designers may use FEMA 452 (FEMA 2005) and FEMA 453 (FEMA 2006). Of particular interest to this Standard is the Building Vulnerability Checklist included in FEMA 452.

Safe havens or "safe rooms" can be designed for a single purpose or multiple purposes. Safe havens provide protection for environmental events such as earthquakes, hurricanes, and tornados. Significant economic gains can be realized by designing safe havens for all known threats (e.g., natural or manmade). Safe havens designed to resist high winds from tornadoes are currently being installed in homes across the United States. Even safe havens for blast loading are not a novel idea. These rooms have been incorporated into construction in the United Kingdom (World War II), the United States (Cold War), and most recently, Israel. Since the Gulf War, the Israeli Defense Force has required apartment protected spaces (APSs) and floor protected spaces (FPSs) to be incorporated into all new construction. For high wind and/or seismic zones, safe havens can be economically incorporated into new construction by utilizing the thick concrete or masonry walls of the safe room as part of the lateral-force-resisting system. Additional information on multihazard design of safe havens can be found in FEMA 453.

Blast events typically occur without warning. The effectiveness of the safe haven for protecting occupants from blast threats, however, is entirely dependent on the amount of warning prior to the event. Even if a safe haven does not provide direct protection for a particular blast event, a safe haven can provide protection until safe evacuation is established following the detonation of an explosive device.

Since most recent aggressor attacks have occurred without warning, the more practical purpose of the safe haven is to protect the occupants of the building from a sequential event that follows an initial attack. Following an explosive detonation, a safe haven provides a protected space in which occupants may gather until law enforcement officials determine that it is safe to evacuate the premises. This is particularly important for terrorist activities in which more than one explosive may be detonated or for situations in which the initial event damages a gas line or other volatile material and causes a secondary explosion.

Uncontrolled fires in target buildings can occur following aggressor attacks. In order to protect the occupants of the safe haven from subsequent fires, the walls, slabs, and roofing of safe havens should be designed for a fire rating that meets or exceeds those used for the building structure's means of egress or primary firewall system.

A safe haven will only be effective if the building in which it is located remains standing. It is often unreasonable to design a safe haven within a building with the expectation that the surrounding structure may collapse. Although the safe haven must be able to resist debris impact, it is not reasonable for it to withstand the weight of the building falling on it, except as required by risk analysis. Therefore, the effectiveness of the safe haven will depend on the ability of the building to sustain damage but remain standing.

When determining the location of a standalone safe haven, the priority is access. In practical terms, standalone safe havens may have little value for use as protection against a blast event due to the time limits typically involved in a blast event. Standalone safe havens have a much greater applicability to some of the other environmental, chemical, biological, or radiological hazards. For a blast event, internal safe havens are generally preferable.

C7.3.2 Applicable Loads and Performance. After the appropriate loads are calculated for the safe haven, they should be applied to the exterior walls, floor, and roof surfaces of the safe haven to determine the design forces for the structural and nonstructural elements. A continuous load path should be provided to carry the loads acting on the safe haven to the ground. Horizontal elements of the safe haven should be designed for uplift forces and load reversals. In order to develop ductile deformations in response to extreme loading, seismic detailing that addresses ductile behavior should be provided in all elements of the safe haven. Design pressure loading shall be based on the threat assessment, with consideration for the loads specified in NFPA 68 (NFPA 2007) or similar requirements specified in the Eurocode 1 (ICAB 2006).

Although a safe haven inside a larger building or otherwise shielded from the hazards is less likely to experience the full effects and missile impacts, it should be designed for the design pressures and potential missile impacts that would apply to a standalone safe haven. There is no conclusive research that can quantify allowable reductions in design forces for safe havens within buildings or otherwise shielded from the hazards. Similarly, the elements of the safe haven should be designed to resist the actual design loads and the code-specified allowable resistance despite the low probability of occurrence of extreme loads.

For design, standard predesigned enclosure and structural elements can be made available that will provide a level of protection for a given class of safe haven. For nonstandard elements, a qualified design professional is required to provide a custom design.

C7.3.4 Location within Building. In-ground safe havens provide the greatest degree of protection against missiles and falling debris. Safe havens located below ground are especially well-protected on the soil sides. Another alternative safe haven location is an interior room on or near the first floor of a building. Multiple safe havens distributed throughout the building provide good access. Safe havens located in the interior of the building may more readily be constructed without windows. The ceiling of the safe haven may consist of hardened sections of the floor structure above the safe haven.

Safe havens located against an exterior wall will likely have glazing fenestration and exterior cladding. Although the general window and cladding system on the building may not have specific resistance to blast, the portion of the cladding enclosing the safe haven should be hardened to provide a high level of protection.

Regardless of where in a building a safe haven is built, the walls and ceiling should protect the occupants from missiles and falling debris; they must remain standing if the building is severely damaged. If sections of the building walls are used as safe haven walls, those sections should be separated from the structure of the building. This is true regardless of whether interior or exterior walls of the building are used as safe haven walls.

While the greater portion of the building may have limited or no protection, building portions adjacent to the safe haven should have some blast-resistant properties. It may be appropriate that all cladding on the floor that contains the safe haven would have a level of protection. The structure above and below the safe haven should have protection to prevent debris from falling on the safe haven and to prevent the support of the safe haven from being lost. This might be required regardless of the design criteria determined for the safe haven itself.

REFERENCES

Federal Emergency Management Agency (FEMA). (2006). Design Guidance for Shelters and Safe Rooms in Buildings, FEMA 453. FEMA, Washington, D.C.

FEMA. (2005). Risk Assessment: A How-To Guide to Mitigate Potential Terrorist Attacks Against Buildings, FEMA 452. FEMA, Washington, D.C.

ICAB SA (ICAB). (2006). Eurocode 1—Actions on Structures, BS EN 1991-1-7:2006, <www.eurocode1.com/en/eurocode1.html> [May 13, 2011].

National Fire Protection Association (NFPA). (2007). Guide for Venting of Deflagrations, NFPA 68. NFPA, Quincy, Mass.

Chapter C8
EXTERIOR ENVELOPE

C8.1 DESIGN INTENT

Failure of the building envelope may result in severe injuries and fatalities. For structures that remain standing in the event of an explosion, the most severe injuries are typically due to hazards caused by failure of the exterior façade. Therefore, it is important to consider the response of the nonstructural façade components, in addition to the structural elements of the building.

Exterior window systems are generally one of the first systems to fail in an explosive event. Window failures for large-scale explosions may occur over a distance measured in miles. Figure C8-1 maps the extent of window breakage and structural damage that resulted from the 1995 bombing attack on the A. P. Murrah Federal Building in Oklahoma City. The explosion at the Murrah Building resulted in glass breakage over a very large area. Several people were killed in nearby buildings by flying debris even though the building structures performed reasonably well. Another example is the bombing of two U.S. embassies in Kenya and Tanzania in 1998. The bombing in Kenya alone resulted in 250 fatalities and more than 5,000 major injuries. Many of these injuries were caused by glass fragments and flying debris.

C8.2 DESIGN PROCEDURES

C8.2.1 General. The resistance-based design approach requires that nonstructural façade components be designed to fully resist blast loads. This is applicable where a high level of protection is desired. For small achievable standoff distances or large design basis threats, the results are generally bunker-like structures with reinforced concrete walls and heavy punched windows.

The hazard-based design approach requires that façade components be designed to reduce potential hazards to building occupants, rather than resisting actual loads due to the explosive threat. Typically, the "blanket" design pressure and impulse[1] may be considerably less than the maximum load applied to the nearest point of a building caused by the curb-side explosive threat. Large parts of the façade may require replacement after an explosion, and a targeted building may experience the potential of serious injury and fatality in the area nearest the explosion. Damage and hazard are less severe as the standoff distance from the explosive source increases. This approach may be deemed acceptable and cost-effective in many cases, given the one-time nature of a blast event and the low probability of such an occurrence for most buildings.

A well-designed system using the hazard-based design approach incorporates balanced design which provides predictable, ductile damage modes while meeting the required level of protection. For a medium level of protection, a triangular load with a maximum design pressure and impulse of 4 psi

(0.028 Mpa) and 28 psi-msec (0.193 MPa-ms) is widely used in practice to mitigate glass hazard. These values are based on an overpressure with no consideration for fragmentation. The hazard-based approach does not consider near-range effects on façade elements.

If design pressures and impulses are adopted that differ from these values, they should be established based on a rational methodology that considers the design basis threat and achievable standoff distances.

C8.2.2 Response Criteria. Several approaches have been developed over the years to define glazing performance levels. Hazard levels are defined by ASTM F1642-04 (ASTM 2010), described in Section 3.3.3 and shown in Fig. C8-2 (ASTM 2009). Another approach that is widely used in the United States was developed by the General Services Administration (GSA) as shown in Fig. C8-3 and Table C8-1 and was adopted by the Interagency Security Committee (ISC) when it developed the ISC Security Criteria. It has therefore been widely used in the development of commercial blast-resistant window systems [(ISC 2004).

The response limits for window frames and mullions in Table 3-2 are based on the values adopted by the DoD (DoD 2009), which are conservative values yet based on the latest available blast test data. Nevertheless, a range of response limits has been published for different levels of protection and various window types. The U.S. Department of State, generally equivalent to a high level of protection, has adopted conservative response limits for blast window steel frames (without muntins) (DoS 2002). The GSA has generally accepted a maximum rotation of 2 deg for a medium level of protection (ISC 2004).

C8.2.3 Analytical Methods. Dynamic methods are recommended for the design of glazing systems for blast effects (Biggs 1964; DoD 2008). Since many typical components are constructed of materials with significant plastic response, the most cost-effective designs are achieved using nonlinear dynamic design approaches. Static design approaches may be used, but typically result in very conservative designs. Static design charts listing dynamic load factors based on the natural period of the system and the shape of the load function are available (DoD 2008).

Blast design of the building envelope aims to model as accurately as possible the system response and predict the subsequent hazards. Damage is acceptable as long as it does not result in serious injury to the occupants for some levels of protection. Therefore, unfactored ultimate strengths of components are generally used in blast design (i.e., no factors of safety). In elastic-static design, using factored strength values translates into a conservative design, but when nonlinear dynamic response is allowed, this may or may not be the case. By underpredicting the strength of one component in a nonlinear design, a designer may inadvertently underpredict the load on a

---

[1] The design load is applied as a "blanket" value to all vertical façade surfaces, including the front, side, and back faces of a building, regardless of their location or orientation relative to the explosive threat.

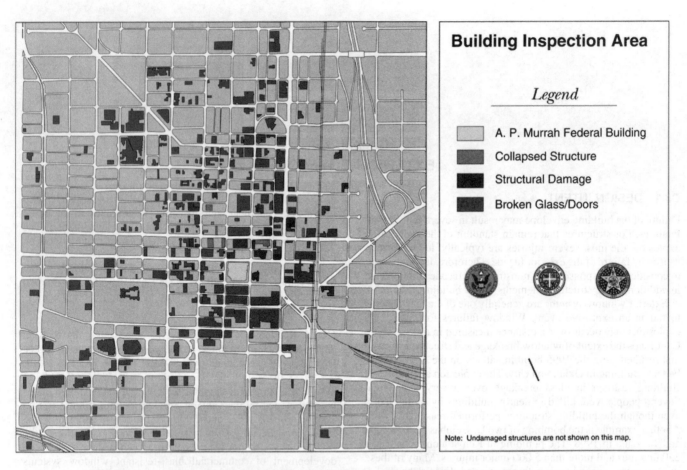

Building Inspection Area

Legend

☐ A. P. Murrah Federal Building

☐ Collapsed Structure

■ Structural Damage

■ Broken Glass/Doors

Note: Undamaged structures are not shown on this map.

FIGURE C8-1. LOCATION OF MURRAH BUILDING AND OTHER DAMAGED STRUCTURES (NFPA 2007).

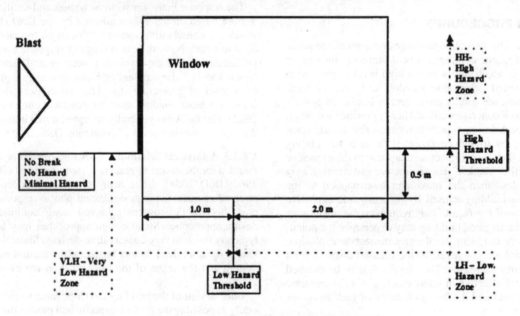

FIGURE C8-2. ASTM F1642-04 HAZARD LEVELS.
Source: ASTM (2010), reprinted with permission, copyright ASTM International, 100 Barr Harbor Drive, West Conshohocken, PA 19428, www.astm.org.

supporting component, thereby underpredicting global damage. If a safety factor is desired for blast design, it should be placed on the loading function or the response limits—not on the material properties.

Average strength factors (ASF) and dynamic increase factors (DIF) are not applicable to glazing.

C8.2.4 Balanced Design. The typical hierarchy of window components, in order of increasing strength, is as follows: glazing, mullions and framing, connections, structural supporting elements (if applicable). It is important to consider the load path and balanced design. Balanced design of a blast-resistant window system translates into a system where the frame and mullions are sufficient to resist the maximum capacity of the glazing, the anchorage is sufficient to resist the maximum loads from the frame and mullions, and the supporting wall system is sufficient to resist the maximum loads developed by the anchorage system.

In conventional construction, the design of façade components is frequently controlled by stiffness rather than strength requirements. To meet the requirement of balanced design for blast, such an approach would result in additional and unnecessary resistance in the components. For example, the connections must be designed to achieve the maximum strength of the supported frame, not the design load.

A significant improvement may be achieved by designing the connections to develop the full static capacity of the associated components. This approach is most applicable in situations where the blast load exceeds the static capacity of the system. For typical loading cases, the dynamic reaction loads at the supports are less than those predicted by a static analysis of the component. The amount depends on the inertia effects in the system. Reaction loads at each end of a simply supported, one-way system are typically between 38% and 50% of the total static resistance of the system even in situations where the actual peak blast loads are significantly higher.

In cases where blast testing is used to verify the performance condition of a design, the requirement for balanced design is dictated by the Authority Having Jurisdiction. This may require further blast testing to failure or additional calculations.

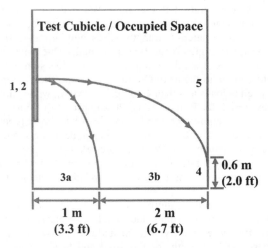

FIGURE C8-3. GSA GLAZING PERFORMANCE CONDITIONS
(Lin et al. 2004).

C8.2.5 Flying Fragments. The design should mitigate, as much as possible, the vulnerability of personnel, structural, and nonstructural components to flying fragments generated by the destruction of "unhardened" (i.e., non-blast-resistant) features such as perimeter walls, adjacent structures, sunscreens, veneers, ornaments, exposed mechanical and electrical equipment, louvers, etc.

C8.3 FENESTRATION

C8.3.1 General. New blast-resistant window systems are discussed in this section. Hazard-mitigating retrofits for window systems are discussed in Section C8.7.

Typically, the following steps are taken when designing new blast-resistant window systems:

1. Size glazing for conventional loads or other special considerations (e.g., forced entry or ballistic resistance).
2. Determine the design blast loads.
3. Modify the glazing as needed to meet the blast requirements.
4. Select a frame that is capable of resisting the estimated dynamic reaction loads from the glazing.
5. Design the anchorage system to develop the full capacity of the glazing and/or framing system.
6. Check the supporting wall capacity to ensure that it can also withstand the blast loads.

The two validation approaches permitted are explosive testing and design calculations. Testing has many advantages since it provides a true representation of a system's performance in an explosive event. However, testing is expensive and, to be effective, every window type and size in the facility needs to be tested. In addition, it is not always easy to replicate final installed building conditions using test structures that are designed to be reusable. Therefore, in many cases engineering calculations are required to ensure that the project performance goals are achieved. One of the most effective approaches is to use testing as a "proof of concept" showing the capability of a given system to meet the performance goals, then to use engineering calculations to adapt that system to various unique conditions found in the building.

C8.3.2 Blast-Mitigating Window Systems
New Systems. The simplest way to minimize the potential glass hazards generated by an explosive event is to develop a complete window system designed to meet the protection requirements. It is much easier to obtain a balanced design when the glazing, frame, anchorage, and supporting wall system are designed as a

TABLE C8-1. GENERAL SERVICES ADMINISTRATION (GSA) GLAZING PERFORMANCE CONDITION

Performance Condition	Protection Level	Hazard Level	Description of Window Glazing Response
1	Safe	None	Glazing does not break. No visible damage to glazing or frame.
2	Very High	None	Glazing cracks but is retained by the frame. Dusting or very small fragments near sill or on floor acceptable.
3a	High	Very Low	Glazing cracks. Fragments enter space and land on floor no further than 3.3 ft (1 m) from the window.
3b	High	Low	Glazing cracks. Fragments enter space and land on floor no further than 10 ft (3 m) from the window.
4	Medium	Medium	Glazing cracks. Fragments enter space and land on floor and impact a vertical witness panel at a distance of no more than 10 ft (3 m) from the window at a height no greater than 2 ft (0.6 m) above the floor.
5	Low	High	Glazing cracks and window system fails catastrophically. Fragments enter space impacting a vertical witness panel at a distance of no more than 10 ft (3 m) from the window at a height greater than 2 ft (0.6 m) above the floor.

Source: ISC (2004).

unit to carry the applied loading within acceptable levels of performance.

A typical blast-resistant glazing system consists of a single laminated pane or an insulating glass unit (IGU) with a sacrificial outer pane and a laminated inner pane. The framing system generally consists of metal elements designed to resist the reaction loads from the glazing. Anchorage of the frame to the walls is a key component in developing a blast-resistant system.

A minimum frame bite is specified for blast windows. Generally, a minimum bite of 0.5 in. (13 mm) can be designed to work for typical commercial applications. Further guidance for determining a minimum bite for laminated glass is given in ASTM F2248-9 (ASTM 2009).

Special Considerations for New Window Systems. For the design of new window systems, the designer has full control over the various components to obtain a balanced and cost-effective design. For example, laminated glass may be more expensive than monolithic glass, but may reduce the glazing reactions so that a lighter framing system is achievable compared to that required for an equivalent monolithic system (monolithic systems are generally thicker and stiffer). In addition, laminated glass performs better in an overload situation compared to monolithic glass, thereby providing an overall reduction of risk to the occupants.

Capabilities of Typical Commercial Window Systems. Conventionally designed windows systems do not generally have the capacity required to resist typical blast loads. For example, a 48-in. by 66-in. (1,219-mm by 1,676-mm) window designed for a 60 psf or 0.42 psi (2.9 kPa) wind load has a total design capacity of approximately 1,320 lb (59 kN). The average design load on the window frame is approximately 5.8 lb/in. (0.66 kN/mm) along the perimeter of the window. Typical peak dynamic reaction loads from a window system designed to achieve a moderate level of protection may be 60 lb/in. (6.8 kN/mm) or more. Even assuming that a fairly high factor of safety was used in the conventional design, the conventional framing system is not capable of resisting reaction loads of this magnitude.

Conventionally designed windows (also known as "punched windows") using 0.25-in. (6-mm)-thick laminated annealed glass may be modified to achieve moderate levels of protection. This may consist of stiffening the existing frame or adding additional anchors. The materials (steel and aluminum) and thicknesses—3/32-in. to 1/8-in. (2 mm to 3 mm)—typically used in conventional punched window frames are capable of achieving reasonable levels of protection. Problems usually occur due to failure of secondary components (snap-in glazing stops, and hinge and latch mechanisms) or due to insufficient anchorage.

Conventional window frames often use snap-in glazing stops. Finite element analysis has shown that typical snap-in glazing stops fail under shear loads as low as 9.0 lb/in. (1.0 kN/mm). This mode of failure may generally be avoided by using glazing stops that are screwed to the frame, designing the glazing stop such that it engages the frame when loaded, or orienting the window such that the glazing stop is on the exterior (this last option may result in rebound failure).

Typical commercial anchorage configurations are usually not sufficient to carry blast loads, but retrofit of the anchorages may increase their capacity. A typical anchorage system consists of self-tapping screws placed at approximately 36 to 48 in. (76 to 102 mm) on center, or anchor straps, generally four to six per window opening. If this type of connection is sufficient for the applied blast loads, no modification of the anchorage is required. However, for most facilities, additional anchorage capacity is required to resist the blast loads. A typical blast design consists of expansion or epoxy anchors at 12 to 16 in. (305 to 406 mm) on center. Table C8-2 shows a sample of typical, commercially available frames and their corresponding capacities.

C8.3.2.1 Glazing. The first step in the design of the window system is to develop a glazing layup that meets the performance requirements of the project. Both the ASTM F1642-04 Hazard Levels (ASTM 2010) and GSA Window Performance Conditions assess the probability of occupant injury by rating the extent to which glass enters the building interior (ASTM 2010; ISC 2004). There are several computer programs available to perform this part of the analysis. Programs such as *SAFEVUE* are available for predicting first crack of glazing systems (NFESC 1995). In addition, programs such as *WINGARD*, *WINLAC*, and *HAZL* are capable of predicting postfailure performance conditions consistent with those used in the ISC Security Criteria (ARA 2005; USACE/ERDC 2000; DoS 1990). Further discussion on the definition and rationale for adopting a maximum probability of failure may be found these documents.

These computer programs perform a nonlinear dynamic analysis and account for numerous failure modes of the glazing, including glass failure, film and laminate failure, and bite failure in a simple standardized fashion using a specified probability of glass failure. The required performance condition for the project should be developed during the project-specific risk assessment. In addition, these software programs may be used to determine the maximum capacity of the glass (also known as "glazing capacity").

Another consideration is that published material properties usually represent the minimum guaranteed strength, not the typical or actual material strength. Glass provides an extreme case of design strengths versus typical strengths. Due to its brittle nature, the break-strength of annealed glass varies significantly from one sample to the next. Out of a given set of test samples, 1% may fail at loads less than 4,300 psi (30 MPa), while another 1% may resist loads exceeding 20,200 psi (140 MPa). The "S-shaped" cumulative probability curves for different types of glass, as generated by the mean and standard deviations in computer programs such as *WINGARD*, are relatively flat for both the "few breaks-per-1,000" strength and for "many breaks-per-1,000" strength. These curves are relatively steep at the mid-range.

Since building codes cannot accept appreciable glass damage in response to extreme winds due to defective glass, they typically specify conservative values for static design—either the "1 break-per-1,000" strength for tempered and heat-strengthened glass or the "8 breaks-per-1,000" strength for annealed glass. However, blast-resistant design guidelines cannot afford to be so conservative. The traditional approach significantly overpredicts the hazards caused by a typical glass window and underpredicts the glazing reaction loads on the window frame.

Therefore, significantly higher probabilities of failure are typically used for blast design. The General Services Administration (GSA) has adopted a conservative probability of failure of "750 breaks-per-1,000" for their *WINGARD* computer program, which is independent of the level of protection (ARA 2005). Earlier, this number was determined by the GSA to best replicate the performance of tested window systems in *WINGARD* and has been used during *WINGARD* validation comparisons on more than 200 explosively tested window systems with reasonable results. The DoD and Department of State have their own recommended criteria for glazing design (DoS 2002; DoD 2002; DoD 2007).

However, the GSA's adopted probability of failure of 750 breaks-per-1,000 has since been recognized as overly conservative by practitioners as the result of additional blast testing. It is

TABLE C8-2. TYPICAL COMMERCIALLY AVAILABLE FRAMING SYSTEM

Description	Figure	Equivalent Static Capacity[a] (Inward)	Possible Dynamic Capacity[b]
Light commercial fixed thermally broken aluminum window system with snap-in glazing stop on interior		Varies, generally controlled by glazing stop, may be as low as 9 lb/in. [1.58E-3 kN/mm]	6.0 lb/in. to 9.0 lb/in. [1.05E-3 kN/mm to 1.58E-3 kN/mm]
Light commercial fixed thermally broken aluminum window system with attached glazing stop on interior		Up to 100 lb/in. [0.018 kN/mm]	67 lb/in. to 100 lb/in. [0.012 kN/mm to 0.018 kN/mm]
Light commercial fixed thermally broken aluminum window system with snap-in glazing stop on exterior		Up to 100 lb/in. [0.018 kN/mm]	67 lb/in. to 100 lb/in. [0.012 kN/mm to 0.018 kN/mm]
Light commercial operable (swing-out) steel framed window system with snap-in glazing stop on exterior		Up to 145 lb/in. [0.025 kN/mm]	97 lb/in. to 145 lb/in. [0.017 kN/mm to 0.025 kN/mm]
Heavy commercial fixed thermally broken aluminum window system with glazing stop angle on interior		Up to 320 lb/in. [0.056 kN/mm]	213 lb/in. to 320 lb/in. [0.037 kN/mm to 0.056 kN/mm]

Description	Figure	Equivalent Static Capacity[a] (Inward)	Possible Dynamic Capacity[b]
Heavy fixed steel window system with glazing stop angle on interior		Up to 440 lb/in. [0.077 kN/mm]	293 lb/in. to 440 lb/in. [0.051 kN/mm to 0.077 kN/mm]
Heavy commercial fixed steel framed window system with tube glazing stop on interior		Up to 750 lb/in. [0.131 kN/mm]	500 lb/in. to 750 lb/in. [0.088 kN/mm to 0.131 kN/mm]

[a]Capacity without modification.
[b]Assumes a dynamic load factor (DLF) of 1.0–1.5.
Source: ARA (2005), with permission.

unreasonable and costly to design a glass make-up using the "750 breaks-per-1,000" strength because this would assume the vast majority of the glass (749 lites) would break before achieving this design loading. Hence, the value has been adjusted by the Committee of this Standard to a more realistic value of 500 breaks per 1,000. Since the cumulative probability curves are fairly steep over this mid-region, the difference between the 500- and 750-breaks-per-1,000 strength is not too great.

For the hazard-based design approach, the ISC Security Criteria specify that at least 90% of the building's glazing area shall also meet the specified performance condition for pressures and impulses resulting from the design basis threat located mid-width of the building at the defended perimeter to further reduce glass hazard (ISC 2004).

C8.3.2.2 Frame and Mullion Design.
A conservative static design for the frame may be conducted using the average peak dynamic reactions from the glazing by distributing the dynamic loads over the perimeter length of the frame. This process is repeated for all four sides of the window system to arrive at a total peak dynamic force applied by the glass to the frame system. This total peak dynamic load is then divided by the area of the window glass to arrive at an average static design pressure. The average static pressure is then used in a traditional static analysis of the frame system.

However, it is usually impractical, or at least uneconomical, to design the window frame to remain completely elastic. Elastic designs for even relatively small explosive devices can be extremely expensive to construct. Therefore, it is important to design the system using ductile components and to account for the dynamic nonlinear response of the system.

For a dynamic analysis, the dynamic reactions from the glazing are applied to a model of the frame system. This results in a less conservative design if nonlinear responses are allowed in the dynamic analysis. These models may be SDOF up to full nonlinear finite element solutions of the frame system. The level of detail required depends on many factors, including the complexity of the frame system, the amount of nonlinear response, and the material makeup of the supporting frame system.

It is important to use the maximum capacity of the glazing system in the design of the window frame. This ensures that the glazing fails before the window frame. For window systems designed to achieve a Performance Condition 1 or 2 response (as defined in the ISC Security Criteria), the glazing may have additional capacity in excess of that required to resist the design load. This additional capacity should be accounted for in the design of the framing system. The maximum capacity of the glazing is used in the design of new window systems and is highly recommended in retrofit designs as well. In retrofit designs where it is not possible to develop the full capacity of the glazing, the framing system should be designed to fail in a controlled manner that achieves the required performance condition. This may involve adjusting the performance condition predicted for the glazing to account for premature failure due to the frame response.

Window mullions are treated similarly to window frames except the mullion generally carries load from two or more panes and is generally attached only at the ends. Mullion response also results in one of two potential modes of failure. The first failure mode, which consists of the mullion failing and detaching from the window system, must be avoided. This can be accomplished by using ductile materials and designing the connections to transfer the full flexural and shear capacity of the mullion to the frame. The second type of failure consists of the mullion deforming in a manner that leads to a premature failure of the glazing. In addition to the glazing loads, a mullion may also carry reactions from attached mullions. The analyst must account for all loads from the glazing and other mullions in the analysis.

C8.3.2.3 Connections and Anchorage. Typical anchorage systems use components that often fail in a brittle manner (e.g., high strength bolts, expansion and epoxy anchors). To determine the adequacy of an anchorage system, the ultimate capacities may be used without typical factors of safety or reductions in load.

The design of the anchorage system depends on the analysis approach used for the framing. If the frame was designed following the static approach, then static reactions from the frame model are applied to the anchorage to determine the appropriate anchorage configuration. The analyst must ensure that all loads required for the anchorage design are included. For example, if a window contains structural mullions, an additional point load from the mullion reaction should be added to the anchorage loads for analysis.

If dynamic analysis is performed, the peak dynamic reaction from the frame and any mullion dynamic loads should be applied to a model of the anchorage. The load function should be based on the time of maximum displacement of the frames and mullions.

Several methods exist for anchoring window systems to the supporting structure. The most common type of anchorage for commercial facilities is expansion or epoxy anchors placed into the concrete or brick supporting walls to carry the blast loads. These anchors may be placed at any required spacing and bolted through the glazing pocket or stop into the supporting wall system. Another common type of window frame anchorage is the use of a steel plate or embed that is cast into the structure and bolted to the frame. These embeds may be either single continuous or multiple independent plates. Steel embeds are generally only used for applications requiring a high level of protection.

C8.3.2.4 Rebound Loads. For systems that fail and enter the occupied space (ISC Performance Conditions greater than 2), the system does not rebound at the design load; therefore, the calculated rebound load equals zero. To obtain a rebound load for design, the analyst should reduce the blast loads until rebound occurs. Generally, the rebound loads occur at 50% to 75% of the inward loading, and a check of the framing and mullions is usually all that is required. If the anchorage is designed for the full inward dynamic loading, it is normally adequate for rebound. Some retrofit designs utilize anchorage arrangements that have different capacities in the inward and rebound directions. In such cases, each direction should be checked.

C8.3.3 Curtain Wall Systems. Storefront and curtain wall system designs vary greatly, and it is difficult to develop generalities about these window systems. However, the glazing for such systems should be designed in a manner similar to that for punched window systems. The major difference is that the glazing may be required to remain in the frame for large curtain wall systems due to the size of the openings or their height above the floor. In addition, the response of large curtain wall systems may be such that the glazing and framing systems cannot be analyzed separately.

In curtain wall systems, as with punched window systems, the anchorage system must be designed to ensure that the frame and/or glazing systems fail before the anchorage. However, the total loads from a curtain wall system are generally much higher and require a more robust structural system to support them. To determine the adequacy of the anchorage system, the ultimate capacities may be used without typical factors of safety or reductions in load.

Depending on the size of the window opening and its attachment to the structure, it is often less expensive or more practical to replace a storefront or curtain wall system than to retrofit existing windows. In addition, for facilities requiring a high level of protection, a replacement window system may be the only viable retrofit.

C8.3.4 Skylights. Glazing for skylights is designed in a manner similar to punched window systems. The major difference is that the skylight glazing is designed to remain in the frame due to the additional hazard of falling glass to occupants. This may require a larger bite, wet-glazing, or an interior catcher system.

C8.3.5 Operable Windows. Operable window frames often have lower capacities than comparable fixed window systems due to failure of the lock and pivot mechanisms. Hazardous failures may be avoided in operable window systems by increasing the number of latches, using stronger locks and hinges, or using swing-out window systems that bear on the subframe when loaded in the inward direction.

C8.3.6 Doors. Design for blast-resistant doors is typically outside of standard practice for a commercial building but may be required in special situations, such as safe havens or as dictated by the project-specific risk assessment. Door frames designed using SDOF methods may adopt the glazing system framing response limits given in Table 3-2.

Exterior public access doors are generally glass, and public emergency egress as well as equipment, maintenance, delivery, and utility doors are generally commercially available hollow metal doors. Doors may be designed to be fully blast-resistant or only to vent blast pressures by opening in the direction of expected blast loads (ASCE 1997). The structure surrounding the blast-resistant door must be capable of resisting the loads transferred from the door. These loads may require substantial structure to allow the doors to reach their design rating.

C8.3.6.2 Public-Access Glass Doors. Glass in public access doors may be designed for blast using approaches similar to those for "punched" windows where the glazing is selected to limit interior debris. For doors required to remain operational in the event of a blast, it is good practice to design the tube sections to remain elastic.

C8.3.6.3 Oversize and Rollup Doors. Oversize and rollup flexible metal doors are often used in delivery and dock areas. These doors have a surprising capacity to resist blast loads because of their inherent ductility. Tests have shown that these doors can resist pressures of up to 3.0 to 4.0 psi (21 to 28 kPa) with no significant inelastic deformations for rotations exceeding 10 degrees. Special track depths should be specified to be at least two times the depth of a non-blast resistant door to ensure that door edges remain in vertical or horizontal tracks.

C8.3.6.4 Other Doors
Egress, Equipment, Maintenance, Delivery, and Utility Doors. Conventional hollow metal or solid core doors are widely used in commercial buildings for emergency egress, equipment access, maintenance access, delivery areas, and utility spaces. Typical sizes include personnel doors, 3 ft × 7 ft (0.9 m × 2.1 m), and equipment or "double" doors, 6 ft × 7 ft (1.8 m × 2.1 m). Hollow metal doors are typically constructed of 16-, 18-, or 20-gauge (1.5-, 1.2-, or 0.9-mm) cold-rolled steel sheets with

typical overall thicknesses of 1–3/8 in. (35 mm) and 1–3/4 in. (44 mm). Most common solid core doors are honeycomb and polystyrene, with or without vertical stiffeners. Most manufacturers dephosphatize and prime door components for corrosion resistance.

Fire rating is a common "discriminator" for the strength of door systems. The most commonly available fire ratings are 3 hours, 1.5 hours, 0.75 hour, and 20 min., and the rating time is generally directly related to the skin thickness of the door; for example, a 1.5-hour fire rating generally is achievable with a 20-gauge (0.9-mm) skin thickness, while a 3-hour rating generally equates to an 18-gauge (1.2-mm) skin thickness.

Door frames for commercial hollow steel and solid core steel sheathed doors are typically installed utilizing two anchorage methods. In structural steel framed buildings, door frames are generally stitch welded to a subframe supported by steel girts. Door frames are installed after walls are complete. In masonry buildings or where masonry infill is used in steel or concrete framed buildings, door frames are generally anchored to the masonry walls with wire anchors during wall construction, where wire ties are extended from the door jamb into the wall as masonry courses are laid. The grouted jamb interlocks the wire anchors, the door frame, and the wall. Masonry anchors are often installed with horizontal reinforcement (typically for every other course of masonry). Frames installed in masonry walls are filled with mortar, making them substantially more rigid and more resistant to rotation.

Conventional hollow metal and solid core metal doors have limited blast capacity primarily due to the "push-through" type of failure observed. Generally, push-through failure occurs at a relatively low blast load because of the lack of structural integrity around the edges of the door. Higher levels of blast load cause the doors to become inoperable, while even higher loads fail all attachments, creating a missile hazard inside the building.

Personnel-size hollow metal and solid core metal doors generally provide a medium level of protection up to static pressures of 1.0 psi (6.9 kPa), but may resist significantly higher pressures for short-duration blasts. Similarly, equipment-size doors gener-

ally provide a medium level of protection up to 0.5 psi (3.4 kPa), but significantly higher for short-duration loads.

Commercial door installations may be retrofitted to minimize debris potential through two approaches. First, door panels may be strengthened with tube sections to add significant blast capacity to the door. The tube sections may provide capacity such that the door panel may be designed in tension membrane between the tube sections. The tube sections act as beams spanning across the short dimension of the door.

A second, more expedient, retrofit technique consists of attaching a cable restraint system to the door to minimize debris should the door fail. Steel cables may provide restraint and may be designed to resist the centrifugal force as the door mass rotates out of the jamb or opening. The ultimate strength of the connections should exceed that of the cables. The cable should be attached as a through-bolt connection near the door centerline with steel plates on both sides of the door. Figure C8-4 illustrates this strengthening concept.

A commercial door test standard is available for testing and specification of blast-resistant hollow metal and solid core metal doors and is referenced in Section 10.6. It is not a dynamic test, but uses calculated equivalent static pressure based on a specified blast pressure and calculated door structural properties. The load is derived assuming elastic response; hence, this is a conservative test. All installations are considered to be three-side-supported unless configured with a structural threshold. The standard requires that sections be fastened to the test frame with the same type, size, and number of fasteners as the actual installation. An air or air-over-water system is employed to produce the uniform load. Four door damage categories are defined in Section 3.3.4.

Commercial Blast Doors. A number of vendors offer door and frame systems that are rated based on blast tests. Typical low-range systems provide protection and functionality up to 3 to 5 psi (21 to 34 kPa) for long-duration loads, and significantly higher pressures for short-duration blast loads. These low-range systems are essentially typical thickness hollow metal doors with thicker "skins" and special internal reinforcement. Thicker

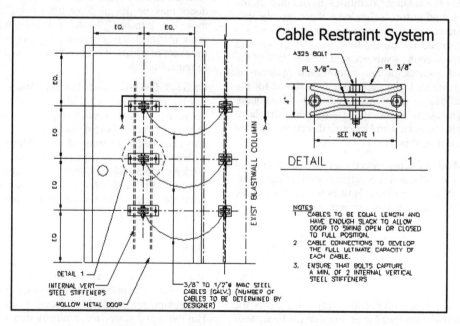

FIGURE C8-4. CABLE RESTRAINT SYSTEM FOR BLAST DOORS.

mid-range systems may provide protection and function up to 20 psi (140 kPa). Special designs may achieve significantly higher pressures. These doors tend to be very heavy and require very special attachment hardware and electromechanical operators to open and close.

C8.4 NON-LOAD-BEARING EXTERIOR WALLS

C8.4.1 General. It may be acceptable for many commercial buildings that are not primary targets for an explosive attack to adopt the hazard-based design approach for exterior walls. As the premium for blast protection is often driven by the cost of the exterior envelope, localized failure of non-load-bearing walls may be acceptable. This eliminates the need to design for the worst-case threat for all exterior walls. For example, the ISC Security Criteria allows the hazard-based approach in some cases, where exterior walls are designed to a nominal (or blanket) pressure and impulse (ISC 2004).

Similar to floor slabs, one-way walls are designed to span in one direction, and two-way walls are designed to span in two perpendicular directions. It is recommended that one-way walls without backing elements span vertically, transferring loads directly into floor diaphragms to reduce the risk of progressive collapse.

C8.4.2 Cast-in Place, Precast, and Tilt-Up Concrete Walls. Concrete construction has a number of advantages with respect to blast: large inertial forces provide favorable response to blast loading; concrete sections are compact, making their behavior fairly predictable; rebound is small due to internal damping; concrete walls provide a good defense against fragmentation; and concrete walls may be designed to meet forced entry and ballistic resistance requirements.

A balance between concrete wall thickness and required reinforcement can be achieved to meet the required level of protection from direct airblast effects and provide shielding from fragments.

Tilt-up construction may be a cost-effective exterior wall solution for blast protection. Depending on the level of protection and design threat, only minimal changes to the tilt panel thickness and reinforcement required for construction may be necessary.

For precast and tilt-up construction, good practice requires that the designer consider pressure leakage between joints as well as connections between the structure and the wall panel.

C8.4.3 Pretensioned and Posttensioned Concrete Wall Panels. Pretensioned and posttensioned wall panels have not been widely used for blast protection, as they typically have insufficient ductility to absorb the energy of an explosion. Load reversals due to rebound and the negative phase of the airblast require that only straight tendons be used.

C8.4.4 Masonry Walls. Reinforced concrete masonry units (CMUs) may provide economical protection for relatively low blast loads. In addition, CMU walls with sufficient thickness and reinforcement may be designed to meet forced entry and ballistic resistance requirements. In general, multiple wythes should be avoided, as continuity between individual wythes is difficult to achieve. Where wall reinforcement is interrupted due to openings, reinforcement in the adjacent wall sections should be modified accordingly to meet the requirements of Chapter 3.

Methods have been established to assess the stability of unreinforced masonry walls under blast loading (DoS 2002; DoD 2008). Note that very slight in-plane vertical movements of "rigid" supports negate the large benefits predicted by assuming arching action.

C8.4.5 Steel Wall Systems. Steel walls have been successfully designed and tested for a range of blast pressures (DoS 2001). As with steel structural elements, non-load-bearing wall components may be designed for large flexural deflections. Steel wall systems generally have the capacity to sustain larger deflections under blast loading compared to more rigid concrete walls. In addition, they achieve greater rebound and should be designed for relatively large stress reversals. Bracing may be required for compression flanges that were formerly in tension. As with concrete, steel connections should be designed such that elements achieve their maximum flexural capacity. For certain types of steel wall construction, the required plate thickness may be dictated by fragmentation protection—making cold-formed steel panels inappropriate. Finally, some types of blast-mitigating steel wall design may be modified to meet forced entry and ballistic resistance requirements.

For unsymmetrical elements, twisting combined with bending occurs when the resultant blast load does not coincide with the shear center. As their behavior is not easy to predict in the inelastic range, the use of unsymmetrical elements should be minimized.

C8.5 ROOF SYSTEMS

Roof components should be designed to force ductile failure modes (e.g., flexure) to precede nonductile failure modes (e.g. shear, punching shear, or connection failure). For threats larger than the design threat, balanced design ensures controlled failure of roof elements. In such a scenario, localized failure of the roof slabs may absorb energy and limit the extent of damage without necessarily causing failure of the roof framing. Failure of the entire roof system is more catastrophic than localized slab failure, as it may lead to progressive collapse.

For steel joists and joist girders of roof systems, the top and bottom chords should be symmetric about the neutral and vertical axes for overpressures and load reversals.

C8.7 HAZARD-MITIGATING RETROFITS

General requirements for blast-mitigating window systems are discussed in Section C8.3.

C8.7.2 Security Window Films. Security window film is frequently used to increase the protection level of existing window systems. These films are adhered to the interior surface of the window to provide fragment retention and reduce the overall velocity of the glass fragments at failure. There are four basic installation methods for security window films. These include daylight installation (applied only to the vision opening), edge-to-edge installation (applied out to the edge of the glass pane), mechanical attachment, and wet-glazed installation (daylight or edge-to-edge film with a silicone bead attaching the film to the supporting frame).

Several different film thicknesses are commercially available. In general, a 0.007-in. (0.18-mm)-thick (or equivalent) film provides adequate performance. Thicker films may be difficult to install in a mechanical application and provide little benefit in daylight applications. These factors should be considered in any upgrade recommendations.

The most common and least expensive application for window retrofit is the daylight application. A daylight film is applied only to the interior vision surface of the glass without attachment to the mullions or frames. Hazard to occupants is reduced with this application, even if the glass–film combination comes out of the frame. It is quickly installed, unobstrusive, and relatively inexpensive. However, at higher pressures, significant hazards and fatalities may result with this retrofit. It is also virtually useless for insulating glass units and degradation due to ultraviolet (UV) light may be lead to reduced performance over time.

A second option is an edge-to-edge application where the film extends into the window frame bite. While this application may provide some additional benefit over a daylight application by exercising the strength of the film membrane in frames with deep glazing pockets, it requires removal of the windows and is more expensive for retrofits. Because of the large increase in cost versus the marginal benefit provided by an edge-to-edge application over a daylight application and the advantages of laminated systems in new applications, the edge-to-edge retrofit has generally been abandoned from consideration by many security planners and blast consultants.

To significantly increase the protection provided by security films, attached systems have been developed. A third option for attaching the film to the window frame is a wet-glazed system. For this application, the film is adhered to the window frame with a large bead of structural silicone. Very limited test data have shown that a properly installed wet-glazed film application provides adequate attachment of the window film to the window frame and provides levels of protection similar to that seen with mechanical attachment (discussed in the next paragraph). This requires the silicone to have adequate overlap onto the window film and the window frame to develop a membrane response in the film. The required overlap depends on the installation. Quality control is essential in a field-applied wet-glazed solution.

For the fourth option, mechanically attached systems use aluminum batten bars that are screwed to the window frame to securely hold the film after glass breakage. The membrane capacity of the film is thereby fully utilized. Several manufacturers have demonstrated with blast testing that mechanically attached window systems provide considerable benefit over unprotected windows and daylight window films.

Fragment-Catching Systems. Fragment-catching systems have also been developed to reduce the potential glazing hazards resulting from an explosive event. Unlike other treatments that are applied directly to the glass (e.g., laminating or attaching a security film), fragment-catching systems are placed behind the window system in an attempt to capture as many glass fragments as possible (Lin et al. 2004). Currently, three basic fragment-catching retrofit systems are readily available: deployable catch bar systems (refer to Section 8.7.4), blast curtain systems (refer to Section 8.7.3), and mesh screens. These systems generally work best when used in conjunction with daylight applied security window film or laminated glazing.

Adequate framing and anchorage must be provided for a deployable catch bar system or mesh screens due to the dynamic loads imparted to the window frame or supporting structure by these types of retrofit systems. If the window frame system or wall system is not capable of supporting the reaction loads imparted by the retrofit, premature failure of the window system will likely occur.

Blast curtains are required to meet performance conditions, rather than prescriptive requirements. This usually involves the need for blast testing of the specific blast curtain type, glazing layup, and window dimensions. A minimal anchorage is required due to the structural response of the blast curtain as it responds to glass fragment hazards. Operable blast curtain system must be closed at the time of the explosive event to be effective. This may affect visibility as occupants must leave blast curtains closed to receive any benefit from the retrofit.

C8.7.5 Secondary Window Systems. Another blast mitigation retrofit consists of installing a secondary blast-resistant window system (also known as "storm windows") behind an unprotected window system. This solution provides an increase in the blast protection without requiring removal or adjustment of the existing window units. It has frequently been used in situations where it is more cost-effective to leave the original windows in place or for historically listed buildings. For example, the architect may desire an exterior look that cannot be economically achieved with a blast-resistant system.

A secondary window retrofit normally consists of a system with a single pane of laminated glass. The framing system generally consists of a metal framing system designed to resist the anticipated reaction loads from the glazing. This solution requires periodic removal or opening of the interior window unit in order to clean the space between the existing and secondary window units. It may require the use of a specially designed window frame that allows access to the enclosed space. Operable windows generally require special consideration in blast design to prevent failure of the latch and pivot mechanisms.

C8.7.8 Fiber-Reinforced Polymers. The U.S. Army Corps of Engineers has successfully developed and tested a retrofit solution using geotexile fabrics for existing reinforced concrete and CMU walls (USACE 1999). Fiber reinforced polymers (FRP), generally in the form of ultra-high-strength carbon or aramid fibers in an epoxy matrix, have been successfully tested for mitigating the effects of blast loading.

C8.8 AMPLIFICATION AND REDUCTION OF BLAST LOADS

C8.8.1 Building Shape and Site. Blast loads applied to the building envelope may be reduced by intelligent selection of the building shape during conceptual design. The reflected pressures applied to the exterior envelope are a function of the angle between the reflected surface and the direction of the shockwave, as well as the number of reflections. Simple convex-, rectangular-, and cylindrical-shaped buildings generally provide better performance than buildings with multiple reentrant corners or overhangs. Concave, L-, and U-shaped buildings may cause multiple reflections from the shockwave, amplifying applied blast loads and increasing damage to the building.

Large fixed objects near a building, such as blast walls, may be taken into account when determining blast loads. However, the advantage of reduced blast loads near the wall is typically offset by reformation of the shockwave a distance away and the uncertainty of the location of the explosive threat.

C8.8.2 Venting. Detailed guidance on the calculation of internal blast loads is given in Chapter 4. In the event of the detonation of a condensed high explosion in a confined structure, two loading phases result (NFPA 2007; DoD 2008). The first phase results from the high pressures generated from the initial shockwave, which are amplified due to multiple reflections. The second phase develops as the gaseous products independently cause the build-up of pressure. For both accidental and intentional explosions, high-risk areas may be designed to reduce confinement by allowing venting of the explosive forces from interior spaces to the building exterior. Locating these areas near the building perimeter facilitates venting and reduces the need for structural hardening.

REFERENCES

American Society of Civil Engineers, Task Committee on Blast-Resistant Design of the Petrochemical Committee of the Energy Division of ASCE (ASCE). (1997). *Design of blast resistant buildings in petrochemical facilities.* ASCE, Reston, Va.

Applied Research Associates (ARA). (2005). *WINGARD,* Version 5.0. Window Glazing Analysis Response & Design User Documentation, ARA-TR-05-15627.5-1 (limited distribution). ARA, Albuquerque, N.M.

ASTM International (ASTM). (2011). Standard Test Method for Metal Doors Used in Blast Resistant Application, F2247-11. ASTM, West Conshohocken, Pa.

ASTM. (2010). "Standard Test Methods for Glazing and Glazing Systems Subject to Air Blast Loadings," ASTM F1642-04 (2010), ASTM, West Conshohocken, Pa.

ASTM. (2009). Standard Practice for Specifying an Equivalent 3-Second Duration Design Loading for Blast Resistant Glazing Fabricated with Laminated Glass, ASTM F2248-09. ASTM, West Conshohocken, Pa.

Biggs, J. M. (1964). *Introduction to structural dynamics.* McGraw-Hill, New York.

Federal Emergency Management Agency (FEMA). (1996). The Oklahoma City Bombing: Improving Building Performance through Multi-Hazard Mitigation (FEMA Bulletin 277). FEMA, Washington, D.C.

Interagency Security Committee (ISC). (2004). ISC Security Criteria for New Federal Office Buildings and Major Modernization (for official use only), <www.dhs.gov/files/committees/gc_1194978268031.shtm> [May 13, 2011].

Lin, L. H., Hinman, E., Stone, H. F., and Roberts, A. M. (2004). "Survey of window retrofit solutions for blast mitigation." *ASCE J. Constr. Fac.,* 18(2), 86–94.

National Fire Protection Association (NFPA). (2007). Guide for Venting of Deflagrations, NFPA 68. NFPA, Quincy, Mass.

Naval Facilities Engineering Service Center (NFESC). (1995). Safety Viewport Analysis Code User Documentation (*SAFEVUE*) Version 2.0. NFESC, Port Hueneme, Calif.

Smith, P. D., and Hetherington, J. G. (1994). *Blast and ballistic loading of structures.* Butterworth Heinemann, Oxford, UK.

U.S. Army Corps of Engineers (USACE). (1999). Airblast Protection Retrofit for Unreinforced Concrete Masonry Walls, ETL 1110-3-494. USACE, Washington, D.C.

U.S. Army Corps of Engineers Engineer Research and Development Center (USACE/ERDC). (2000). *HAZL version 1.0 manual.* Vicksburg, Miss. U.S. Department of State (DoS). (2002). *A&E design guidelines for U.S. diplomatic mission buildings.* U.S. Department of State, Overseas Building Operations, Washington, D.C.

U.S. Department of Defense (DoD). (2009). *Blast, Ballistic, and Forced Entry Resistant Windows,* UFC 4-023-04.

DoD. (2008). Structures to Resist the Effects of Accidental Explosions, UFC 3-340-02, <www.wbdg.org/ccb/DOD/UFC/ufc_3_340_02.pdf> [May 12, 2011].

DoD. (2007). DoD Minimum Antiterrorism Standards for Buildings, UFC 4-010-01, <http://www.wbdg.org/ccb/DOD/UFC/ufc_4_010_01.pdf> [May 12, 2011].

DoD. (2002). Design and Analysis of Hardened Structures to Conventional Weapons Effects, UFC 3-340-01 (for official use only).

U.S. Department of State (DoS). (2001). The Steel Stud Wall/Window Retrofit: A Blast Mitigating Construction System, DS/PSD/SDI—TIB #01.01. U.S. Department of State, Bureau of Diplomatic Security, Washington, D.C.

DoS. (1990). WINdow Lite Analysis Code (*WINLAC*) User Documentation (limited distribution). U.S. Department of State, Office of Foreign Buildings Operations, Sverdup-Embassy Task Group, Washington, D.C.

Chapter C9
MATERIALS DETAILING

C9.1.1 Scope. This chapter lists the requirements for detailing of structures to provide levels of protection (LOPs) I through IV. Rational analysis may be used to evaluate LOP compliance of existing structures in conformance with the provisions of Chapters 3 and 9.

While some provisions in this chapter are related to progressive collapse resistance, this chapter is not intended to supplant established design procedures to mitigate progressive collapse.

This standard achieves levels of blast resistance by two independent methods:

1. Minimum detailing requirements
2. Maximum response limits.

Minimum detailing requirements help ensure that the structure has sufficient toughness to meet the performance expectations defined in Chapter 3. Application of the minimum requirements in this chapter is not necessarily sufficient to achieve the intended LOP for a specified threat. Element-specific structural design must be conducted to completely size and detail elements to provide the intended performance.

Minimum detailing requirements are typically the same for all levels of protection. At the lowest LOP, minimum detailing is required to achieve the expected high element deformations and provide associated energy absorption. At higher LOPs, while permanent deformations are expected to be negligible, the minimum detailing provides increased margin against uncertainty, commensurate with the higher LOP expected from the structural system.

Maximum response limits vary significantly between LOPs. Member sizing and structural system design will require corresponding variation. Higher LOPs will dictate structural elements with higher capacities. Likewise, smaller strength is associated with lower LOPs. That is, demand/capacity ratios for (1) LOP I components will exceed 1.0 by a substantial margin, and (2) LOP IV components will be approximately 1.0.

The response limits in Tables 3-2 and 3-3 are calibrated to member responses determined by single-degree-of-freedom (SDOF) modeling (as discussed in Section C.3.4). Rational analysis can be used to map SDOF deformations to the physical system. Such methods allow Tables 3-2 and 3-3 to be used as a guide for member responses calculated by means other than SDOF modeling. Qualitative considerations of member responses are provided in the Commentary for a number of material systems, which provide a framework for assessing member conformance to LOPs when not using SDOF-based design.

The combination of generally consistent structural toughness and scalable capacity provides a means for achieving a balance between design economy and protection.

C9.1.2 Structural Interaction
C9.1.2.1 Primary and Secondary Elements. Elements designated as secondary when considered independently may need to be designed as primary members if necessary to support the action of other primary members. For example, purlins may need to be designed as primary members if they are critical to the lateral bracing of a girder.

C9.1.2.2 Predominance of Global Structural Performance. It is possible that the performance intent of different elements may be in conflict. In any such situation, design decisions will be made that best satisfy global structural performance. For example, detailing to develop tensile membrane action could create a transfer of load that would cause disproportionate damage to the surrounding structure if not properly equilibrated.

C9.1.3 Materials. Materials other than those specified in this chapter may be used if shown by testing or rational analysis to satisfy the performance requirements of this Standard.

C9.1.7 Use of Reference Documents. Chapter 9 provides minimum detailing requirements. These requirements are not necessarily sufficient to realize the intended LOP. For example, not all materials sections in this chapter provide minimum detailing requirements for direct shear affects. This does not imply that direct shear does not need be addressed. It is the responsibility of the registered design professional to determine that the design and detail of the blast-resistant construction meets the LOP.

C9.1.8 Special Inspection. Recommended practice is to conduct special inspections per Chapter 17 of the International Building Code (IBC; ICC 2006), regardless of seismic design category (SDC).

C9.2 CONCRETE

C9.2.1 Scope. Qualitative damage expectations for reinforced concrete are provided in Table C9-1. ACI 318-08 (ACI 2008) serves as the reference document for reinforced concrete. Herein, substantial use is made of the seismic design and detailing requirements of ACI 318-08, because seismic design of conventional buildings assumes significant inelastic response. Although the inelastic demands are quite different in components of reinforced concrete buildings subjected to earthquake and blast loadings, there is no better source of prescriptive details for inelastic response of reinforced concrete components than ACI 318-08.

C9.2.2 General Reinforced Concrete Detailing Requirements
C9.2.2.1 The use of a minimum compressive strength of 3,000 psi (20.7 MPa) mirrors the requirements of ACI 318-08 (ACI 2008) for earthquake-resistant framing.

C9.2.2.2 Only normal-weight concrete components can be designed and detailed using the procedures of Section 9.2. Explicit rules have not been written for the use of light- or heavy-weight concrete because most of the test specimens that form the engineering basis for blast-resistant design have been constructed with normal-weight concrete.

TABLE C9-1. Qualitative Damage Expectations for Reinforced Concrete Elements

Element	Limit State	Superficial	Moderate	Heavy	Hazardous
Beam and column	Reinforcement	No damage	No damage	Local buckling of longitudinal reinforcement	Fracture of longitudinal and transverse reinforcement
Beam and column	Core concrete	No visible, permanent structural damage	Minor cracking (repairable by injection grouting)	Substantial damage	Rubble
Beam and column	Cover	No visible, permanent structural damage	Substantial spalling	Lost	Lost
Beam and column	Stability	None	None	Local buckling of longitudinal reinforcement	Global buckling
Connections	Reinforcement	None	None	Limited fracture and compromised anchorage at joint (load transfer maintained)	Fracture and loss of anchorage at joint
Connections	Concrete	No visible, permanent structural damage	Minor spalling and cracking (repairable)	Substantial damage	Rubble at core
Slab	Diaphragm	Hairline cracking in the vicinity of the blast. Diaphragm action uncompromised for the lateral force and gravity force resistance.	Spalling of concrete cover unlikely except in the immediate vicinity of blast. Connection to supporting beam intact except in the immediate vicinity of blast where localized separation is likely. Diaphragm action uncompromised for lateral force and gravity force resistance.	Minor damage to concrete and reinforcement. Connection to supporting beam yields but fracture is likely in vicinity of blast resulting in localized separation.	Significant damage to concrete and reinforcement. Diaphragm action compromised for lateral force resistance but provides stability for gravity force resistance.

C9.2.2.3 ASTM A706-06a (ASTM 2006) reinforcement is specified for components that might undergo significant inelastic action because such reinforcement has well-defined mechanical properties, a large elongation at break, and is weldable.

C9.2.2.4 Lap splices are permitted in UFC 3-340-02 (DoD 2008). The limits on the percentage of bars that can be spliced at a given location and the distance between splices zones are based on the judgment of the authors of this ASCE Standard and not test data. It is good practice to limit or restrict the use of lap splices in expected plastic hinge zones.

C9.2.2.6 This rule represents a departure from the restrictions of UFC 3-340-02 (DoD 2008), which require demonstration of both tensile strength and ductility of the splice. The UFC guidance is based on tests conducted in the 1960s–1970s on mechanical splices that are different from those currently used. UFC 3-340-02 requires testing of mechanical splices at high strain rates, which are required to support their use at any location in a structural element. Such tests must demonstrate that the coupler is stronger than the joined bars, namely, failure of the test specimen is remote from the coupler. UFC 3-340-02 assumes strain rates of 0.1 in./in./sec and 0.3 in./in./sec (0.1 mm/mm/s and 0.3 mm/mm/s) for far and close-in design ranges, respectively. Strain rates as high as 3.25 in./in./sec (3.25 mm/mm/s) are reported in the literature (Zehrt and Lahoud 1994; Flathau 1971). It is good practice to limit or restrict the use of even ASTM A706-06a (ASTM 2006) splices in expected plastic zones.

C9.2.2.7 This rule represents a departure from the restrictions of UFC 3-340-02 (DoD 2008), which require demonstration of both tensile strength and ductility of the splice. UFC 3-340-02 requires testing of welded splices at high strain rates, which are required to support their use at any location in a structural element. Such tests must demonstrate that the splice is stronger than the joined bars, namely, failure of the test specimen is remote from the welded splice. UFC 3-340-02 assumes strain rates of 0.1 in./in./sec and 0.3 in./in./sec (0.1 mm/mm/s and 0.3 mm/mm/s) and for far and close-in design ranges, respec-

tively. Strain rates as high as 3.25 in./in./sec (3.25 mm/mm/s) are reported in the literature (Zerht and Lahoud 1994; Flathau 1971).

C9.2.2.8 The definitions of columns and beams are taken from Section 21 of ACI 318-08 (ACI 2008).

C9.2.3 Columns

C9.2.3.1 The limits on longitudinal reinforcement ratios are taken from Section 21.6.3.1 of ACI 318-08 (ACI 2008); see that Commentary (ACI 318R-08; ACI 2008) for information.

C9.2.3.2 The requirements for transverse reinforcement are taken from Sections 21.6.4.1, 21.6.4.2, and 21.6.4.3 of ACI 318-08 (ACI 2008). Transverse reinforcement is required over the clear height of the column because the locations of inelastic action are unknown.

C9.2.3.3 The design rule for column shear force is taken from Section 21.6.5.1 of ACI 318-08 (ACI 2008); see that Commentary (ACI 318R-08; ACI 2008) for information. Additionally, it is at the discretion of the design professional to determine whether the shear force should be based on the maximum loading condition or the maximum resistance attained by the framing elements (consistent with balanced design). Herein, transverse reinforcement for shear can be proportioned assuming a contribution from the concrete since the core concrete will be well confined per Section 9.2.5.2 of this Standard.

C9.2.4 Beams

C9.2.4.1 The clear span limit for beams in special moment-resisting space frames can be found in Section 21 of ACI 318-08 (ACI 2008); see that Commentary (ACI 318R-08; ACI 2008) for information. Beams that do not satisfy this criterion should be designed with specific attention to shear capacity using methods provided in other sources such as UFC 3-340-02 (DoD 2008).

C9.2.4.2 The longitudinal reinforcement limits are taken from Sections 21.5.2.1 and 21.5.2.2 of ACI 318-08 (ACI 2008); see that Commentary (ACI 318R-08; ACI 2008) for information.

C9.2.4.3 Confinement reinforcement is required over the clear span of the beam because the locations of inelastic action are unknown. Seismic hoops are required to provide confinement.

C9.2.4.4 The design rule for beam shear force is taken from Section 21.5.4.1 of ACI 318-08 (ACI 2008); see that Commentary (ACI 318R-08; ACI 2008) for information. Herein, transverse reinforcement for shear can be proportioned assuming a contribution from the concrete since the core concrete will be well confined per Section 9.2.5.2 of this Standard.

C9.2.5 Beam-Column Joints

C9.2.5.1 The basis for this rule is Section 21 of ACI 318-08 (ACI 2008). The maximum force to be considered in the design of a beam-column joint should account for the expected yield strength, some strain hardening of the reinforcement, and an increase in strength due to dynamic loading in the beam reinforcement.

C9.2.5.2 The limits set forth in this section are taken from Sections 21.7.2.2, 21.7.2.3, and 21.7.5 of ACI 318-08 (ACI 2008); see that Commentary (ACI 318R-08; ACI 2008) for information.

C9.2.5.3 The basis for this rule is Section 21 of ACI 318-08 (ACI 2008). No reduction in the volume of transverse reinforcement in the joint is permitted as a function of the framing surrounding the joint because the framing might be destroyed by the blast loading.

C9.2.5.4 The computation of joint shear strength is based on Section 21.7.4 of ACI 318-08 (ACI 2008); see that Commentary (ACI 318R-08; ACI 2008) for details.

C9.2.6 Slabs

C9.2.6.1 Limits are set on the minimum and maximum reinforcement ratios to comply with ACI 318-08 (ACI 2008) lower bound limits and to provide large ductility, which is achieved with small reinforcement ratios in slabs.

C9.2.6.2 Symmetrical flexural reinforcement must be provided in the top and bottom of slabs thicker than 6 in. (150 mm) to account for rebound. See Chapter 6 for additional commentary.

C9.2.6.3 This provision is intended to provide some resistance to progressive collapse that might result from the loss of interior support.

C9.2.6.4 This provision is intended to provide some resistance to progressive collapse that might result from the loss of interior support.

C9.2.6.5 Per this provision, the required shear strength is based on the flexural strength of the member rather than calculated reactions, with the intent to provide sufficient shear capacity to resist the maximum possible reaction from flexural response of the element.

C9.2.7 Walls

C9.2.7.1 For walls providing gravity load support to beams, columns, and slabs, it is recommended that they be constructed with columns at each end of the wall and a beam at each floor level that spans between the columns. This practice will provide a gravity load path in the event that the wall, or part thereof, is damaged by the blast loading. Columns shall comply with Section 9.2.3 of this Standard, beams shall comply with Section 9.2.4, and beam-column joints shall comply with Section 9.2.5.

C9.2.7.2 The basis for this rule is Section 21 of ACI 318-08 (ACI 2008) for special structural walls; see that Commentary (ACI 318R-08; ACI 2008) for information. "Wall web" refers to the region of the wall away from edge anchorage, which is also the region of the wall responding principally in flexure under blast load effects.

C9.2.7.3 The diagonal tension rules and diagonal compression limits of Section 21 of ACI 318-08 (ACI 2008) apply to the in-plane shear strength of structural walls. The shear stress limit is that of ACI 318-08 for squat reinforced concrete walls with the capacity reduction factor set to 1.0. For out-of-plane shear forces, a wall should be checked as a one- or two-way spanning slab per ACI 318-08.

C9.2.8 Tension Ties

C9.2.8.2 The tie reinforcement will not likely be exposed to forces resulting from direct airblast, so the strength increases associated with high strain rate loading are not permitted for the calculation of reinforcement strength.

C9.2.8.3 Splices of tension tie reinforcement shall develop the nominal tensile strength of the bar.

C9.3 STRUCTURAL STEEL

C9.3.1 Scope. Qualitative damage expectations for structural steel are provided in Table C9-2.

C9.3.2 General Structural Steel Requirements

C9.3.2.4 Where moment connections are used, the provisions of ANSI/AISC 360-10 Chapter 9 (AISC 2010) should also be considered, particularly Sections 9.3–9.7, which include provisions for panel zone reinforcement, continuity plates, strong-column weak-beam design, and lateral bracing.

C9.3.2.6 The average strength factor (ASF) of 1.1 (Chapter 3) is appropriate for wide-flange shapes. For other member types, the designer may choose to apply R_y and R_t from Section 6 of ANSI/AISC 360-10 (AISC 2010) to F_y and F_u, respectively. These values are typically greater than 1.1. R_y and R_t should not be used in addition to the ASF.

TABLE C9-2. QUALITATIVE DAMAGE EXPECTATIONS FOR STRUCTURAL STEEL ELEMENTS

Element	Limit State	Superficial	Moderate	Heavy	Hazardous
Beam and column	Hinge	Member remains essentially elastic	Yield strain exceeded at extreme fibers. No hinges.	Full plastic hinge formation. Moderate rotations.	Full plastic hinge formation. Extreme rotations.
Beam and column	Stability	Member remains essentially elastic	Small permanent deflection. No instability for primary members.	Local buckling	Local and global buckling
Connections		No visible, permanent structural damage	Connections may exhibit partial yield but should not have lost any capacity	Primary connections such as splices, base plate (if above slab), and beam-to-column connections shall not fail	Connection failure possible, but not before developing ductile element response for beams and columns

Element	Limit State	Superficial	Moderate	Heavy	Hazardous
Concrete slab on metal deck	Flexure	Deformation or yielding unlikely in the vicinity of the blast. Hairline cracking in the vicinity of the blast. Tension reinforcing steel below yield strain. No spalling of concrete. Deck distortion and separation from supporting beams unlikely in the vicinity of the blast. Deflections above allowable are unlikely in the vicinity of the blast.	No permanent deformation or yielding. Slab cracks are repairable except in the immediate vicinity of blast. Applied moment is limited to yielding of tension reinforcing steel. Spalling of concrete cover unlikely except in the immediate vicinity of blast. Metal deck distortion limited to immediate vicinity of blast. Connection to supporting beam intact except in the immediate vicinity of blast where localized separation is likely. Permanent deflections are below allowable. Partial collapse is not permitted in the vicinity of the blast.	Slab yields in flexure. Slab cracks are unrepairable but do not compromise load carrying capacity. Applied moment results in yielding of tension reinforcing steel but does not exceed M_n (concrete strain less than 0.003). Minor concrete spalling of reinforcing steel cover occurs due to impact. Metal deck yields but not fractured. Connection to supporting beam yields but fracture is likely in vicinity of blast resulting in localized separation. Permanent deflections below six times allowable.	Slab develops clear yield lines. Concrete cracks through the entire cross section due to uplift force. Applied moment exceeds M_n resulting in concrete strains in excess of 0.003 and rupture or buckling of reinforcing steel. Concrete crushes resulting in visible spalling and exposing top reinforcing bars due to impact. Metal deck connection to supporting beam fractured and resulted in complete separation. Permanent deflections in excess of six times allowable. Catenary action compromised. Sections of slab hinge or fall onto floor below but without significant impact.
Concrete slab on metal deck	Diaphragm	Diaphragm action uncompromised for lateral force and gravity force resistance	Diaphragm action uncompromised for lateral force and gravity force resistance	Catenary action uncompromised with no sections of slab collapsing. Localized damage to the diaphragm. Diaphragm action uncompromised for lateral load and gravity load resistance.	Diaphragm action compromised for lateral force resistance but provides stability for gravity force resistance. Limit major structural damage to immediate vicinity of blast, not exceeding 50% of the effective diaphragm depth.

C9.4 STEEL/CONCRETE COMPOSITE

C9.4.1 Scope. Qualitative damage expectations for steel/concrete composite elements are provided in Table C9-3.

C9.4.2 Concrete Slab on Metal Deck. In structural steel-framed building construction, the concrete floor slab is typically supported on a cold-formed corrugated metal deck. A noncomposite floor deck only provides a form for the structural concrete slab, whereas a composite floor deck has a shape allowing it to interlock with the hardened concrete, which acts as a permanent form and as the positive flexural reinforcement for the structural concrete slab. In most cases, it is not practical or possible to design the system to maintain composite action between the deck and the slab under blast loading. Therefore, the inclusion of composite action in the analysis of the composite deck should be carefully considered.

C9.4.2.3 In some instances, it might be possible to demonstrate that a light-weight concrete slab is adequate when its function for blast resistance is limited to diaphragm action. If the slab is required to provide flexural resistance to blast loads (particularly those applied directly to the slab), normal-weight concrete should be used.

C9.4.2.4 Wire reinforcement should be consistent with provisions from the Wire Reinforcement Institute (Hartford, Conn.). See the Wire Reinforcement Institute's *Manual of Standard Practice—Structural Welded Reinforcement*, WWR-500 (WRI 2006a) and *Structural Detailing Manual*, WWR-600 (WRI 2006b).

There are no established data or guidance on the performance of welded wire reinforcement (WWR) under blast load conditions, specifically with regard to ductility and high strain rate. WWR has been historically viewed as less ductile than ASTM A615-09b (ASTM 2009) or ASTM A703 (ASTM 2006) reinforcing steel. However, this may be at least in part an artifact of historical WWR behavior, before refinements in the manufacturing process. Current WWR elongation data can be found in WWR-500, p. 9, Table 3(b) (WRI 2006a). (For example, total elongation for wire sizes ranging from W3 to D12 ranges from 7.2% to 13.4%.)

Further information on WWR ductility can be found in ACI Journal Technical Paper Title No. 80-41 (Dove 1983). This article notes that the standard 10-in. (250-mm) gage length tensile specimen does not accurately reflect the ductility of cold-drawn wire.

Since no high-strain-rate tests have been performed on WWR to evaluate its performance under blast loading, UFC 3-340-01 (DoD 2002), UFC 3-340-02 (DoD 2008), and DOE/TIC-11268 (DoE 1992) do not consider or allow its use in blast-resistant construction. WWR is found in other blast-resistant design guidance, but not in great detail. The U.S. Army Corps of Engineers Protective Design Center's SBEDS manual (PDC-TR 06-01 and PDC-TR 06-02; USACE 2006), for example, does include WWR in the reinforcing menu, but does not provide response limits for WWR as it does for reinforcing steel.

In the absence of established guidance, a rotation limit of 2 deg would result in elongations with substantial margin to the published ultimate elongations for typical slab applications. Care should be used in applying such a limit. It does not necessarily imply that WWR should thus be restricted to LOP III and IV (which, as the highest levels of protection, have the lowest target response levels). Other appropriate applications could include an exterior blast, in which the interior pressurization and floor slab response are expected to be less severe than for the exterior structural system.

C9.4.2.9 This provision addresses vertical restraint of the slab, not resistance of horizontal shear.

C9.4.2.11 This provision addresses vertical restraint of the slab, not resistance of horizontal shear.

C9.5 MASONRY

C9.5.1 Scope. Qualitative damage expectations for steel/concrete composite elements are provided in Table C9-4.

TABLE C9-4. QUALITATIVE DAMAGE EXPECTATIONS FOR MASONRY

Element	Limit State	Superficial	Moderate	Heavy	Hazardous
Reinforced	Multiple	Masonry is effective in resisting moment and shear. The cover over the reinforcement on both surfaces of the wall remains intact. System should essentially remain elastic. Axial load-bearing elements maintain bearing capacity. No connection or shear failure. Compressive strain in masonry does not exceed 0.003.	Masonry is crushed and not effective in resisting moment; however, shear resistance remains though interlocking. Compression reinforcement equal to the tension reinforcement is required to resist moment. The cover over the reinforcement on both surfaces of the wall remains intact. No spalling or scabbing of masonry. Axial load-bearing elements maintain bearing capacity. No connection or shear failure. Compressive strain in masonry shall not exceed 0.007.	The concrete cover over the reinforcement on both surfaces of the element is completely disengaged. Equal tension and compression reinforcement that is properly tied together is required to resist moment. Limited spalling and scabbing of masonry. Axial load-bearing elements maintain bearing capacity. No connection or shear failure. Compressive strain in masonry shall not exceed 0.010.	Masonry is completely disengaged. Significant spalling and scabbing of masonry. Axial load-bearing elements maintain bearing capacity. No connection or shear failure. Compressive strain in masonry exceeds 0.010.
Unreinforced	Multiple	Not permitted	Not permitted	Significant spalling and scabbing of masonry. No connection or shear failure. Compressive strain in masonry does not exceed 0.003.	Significant spalling and scabbing of masonry. No connection or shear failure. Compressive strain in masonry does not exceed 0.007.

C9.5.2 General Design Requirements. Reinforced masonry subjected to service design loads (except airblast) can be designed using both allowable and strength design methodologies. For airblast loads, however, only strength design methodology shall be used. Reinforced masonry elements subjected to airblast loads are likely to experience large nonlinear deformations with relatively large strain and ductility demands. Therefore, it is necessary to employ design methodology which can account for response well beyond linear limit states. Working stress design, although much simpler than strength design, only accounts for response below the nonlinear range, and implies that all reinforced masonry shall be designed linearly. Furthermore, since airblast design requires that structural elements respond in ductile manner with large nonlinear deformations, and since allowable design procedure is incapable of predicting response beyond the linear range, allowable stress design procedure is prohibited for airblast design of reinforced masonry.

C9.5.4.4 Generally, airblast load is applied over very short duration, and it is expected that reinforcing bar yielding will occur with high strain rate. The effects of high strain rate on mechanical splices are not well studied, and therefore it is recommended that the use of mechanical splices be limited to zones that remain elastic during the airblast loading response. Furthermore, regardless of the analysis results, it should be conservatively assumed that reinforcing bar will yield, and therefore mechanical splices should develop full (ultimate) dynamic design strength of connected reinforcing bars. That is, if the reinforcing bar reaches yield stress, yielding should occur in the bar and not in the splice (or near the splice). If mechanical splices are placed in zones that are expected to undergo inelastic response, it should be demonstrated by testing that the splice develops full (ultimate) dynamic design strength of connected bars, and that splice sleeve itself does not compromise the integrity of fully grouted masonry. Testing should resemble the actual airblast loading condition as closely as possible.

C9.5.4.5 (also see C9.5.4.4) The effects of high strain rates on welded reinforcing bars, including heating effects, are not well studied, and therefore the application of welded splices should be limited to zones that remain elastic during the airblast loading response. Furthermore, and regardless of the analysis results, it should be conservatively assumed that reinforcing bar will yield, and therefore welds should develop full tensile (ultimate) strength of the connected reinforcing bars. That is, if the reinforcing bar reaches yield stress, yielding should occur in the bar and not in the splice (or near the splice). If welded splices are placed in zones that are expected to undergo inelastic response, it should be demonstrated by testing that the splice develops full tensile (ultimate) strength of connected bars and that it does not compromise the integrity of fully grouted masonry. Testing should resemble the actual airblast loading condition as closely as possible.

C9.5.5 The general intent for wall detailing is to provide sufficient shear resistance to develop the full flexural capacity of the wall.

C9.6 FIBER REINFORCED POLYMER (FRP) COMPOSITE MATERIALS

C9.6.1 Scope
FRP Systems. Two main categories of FRP applications may be identified with respect to explosive blast-resistant FRP design, namely:

1. Structural elements where the blast load is carried by an element made entirely of FRP matrix composite materials. Examples of such FRP construction include FRP blast walls, self-contained protective FRP blast rooms, and FRP blast partitions, all of which prevent (a) the penetration of fragmentation threats and other debris; (b) the formation of spallation; and (c) wall panel rupture due to explosive blast overpressure.

2. Retrofit blast-hardening FRP applications, where FRP construction in the form of an add-on laminate or fabric wrap, are used to strengthen or otherwise reinforce an existing primary load-bearing steel, concrete, aluminum, or wooden structural element. The add-on FRP material serves to increase existing structural element strength and/or ductility, where weight and space are significant constraints.

TABLE C9-5. FIBER REINFORCED POLYMER MATRIX COMPOSITE FAILURE MODES

Fiber-Dominated	Resin Matrix-Dominated	Interface-Dominated
Fiber pull-out	Transverse cracking	Interface debonding
Fiber tensile failure	Interlaminar cracking	Interface delamination
Fiber microbuckling	Intralaminar cracking	Compressive delamination
Fiber shear failure	Edge delamination	

TABLE C9-6. QUALITATIVE DAMAGE EXPECTATIONS FOR FRP-STRENGTHENED, REINFORCED CONCRETE BEAMS AND SLABS

Mechanism	Superficial	Moderate	Heavy	Hazardous
Detachment at substrate	No	Limited	Yes	Yes
Detachment at FRP	No	Limited	Yes	Yes
FRP ply delamination	No	Limited	Yes	Yes
FRP ply edge delamination (peeling)	No	Limited	Yes	Yes
Fiber rupture of breakage	No	Limited	Yes	Yes
FRP lap splice failure	No	Limited	Yes	Yes
Matrix cracking	No	Limited	Yes	Yes
Matrix crazing	No	Limited	Yes	Yes

FRP Durability. The cured FRP laminate should exhibit a void content no greater than 0.5% by volume in order to avoid moisture intrusion or capillary action, resulting in the degradation of the fiber/resin interfacial chemical bond by intrusion of water at the fiber/resin interface. In addition, the resin matrix should exhibit resistance to chemical, solvent, and thermal degradation as required by the operating environment. The resin matrix should exhibit a glass transition temperature, T_g, (as determined by dynamic mechanical analysis) such that the FRP system T_g be 50 °F (10 °C) greater than the maximum operating service temperature, in order to avoid laminate deformation and distortion at elevated temperatures. An FRP system should be selected that has undergone durability testing consistent with the application environment. Durability testing may include hot-wet cycling, freeze-thaw cycling, alkaline immersion, ultraviolet exposure, etc. FRP material selection should follow the durability requirements of *MIL Handbook 17*, Volume 3 (ASTM 2002c).

FRP Failure Criteria. There are a multitude of failure modes associated with FRP composites. From first ply failure to last ply failure, as many as 11 different failure modes (*MIL Handbook 17*, Volume 3, Chapter 5; ASTM 2002c) may be invoked prior to achieving the ultimate load-carrying capacity of the FRP laminate. Table C9-5 identifies these failure modes as falling into three distinct categories: fiber-dominated, matrix-dominated, or interface-dominated failure modes. Each failure mode absorbs its own increment of energy, resulting in a collective total energy absorption that exceeds the shear distortion energy associated with isotropic metals, when compared on an equivalent weight basis.

C9.6.3 General. Data, testing and sampling guidance that can be used to determine material specific strength increase factors can be found in Bank et al. (2003), Zuriek et al. (2006), and Sierakowski et al. (1997).

C9.6.3.1 FRP Delamination Due to Stress Wave Propagation. This requirement provides a minimum tensile strength equated to exhibiting the necessary strain rate response to survive explosive blast strain rate conditions at scaled distances of $Z \geq 1.0$.

A common failure mode associated with FRP laminate construction subjected to explosive blast overpressure is matrix-dominated ply delamination. The FRP laminate is subjected to stress wave propagation through the thickness of the laminate. As the overpressure contacts the outer surface of the FRP laminate, a compressive stress wave is developed, which becomes a tensile stress wave upon reflection from the back face of the laminate which is bonded (i.e., anchored) to a different material substrate (e.g., reinforced concrete, steel, aluminum). If the magnitude of the reflected (i.e., tensile) stress wave exceeds the ultimate tensile strength of the FRP laminate in the through-thickness direction, then ply delamination will occur.

In order to avoid FRP delamination due to blast overpressure-induced stress wave propagation (i.e., stress wave reflection), the explosive blast-reflected pressure magnitude impinging on the FRP laminate surface should not exceed 80% of the ultimate tensile strength of the neat (i.e., unreinforced) resin matrix. Such a condition may be satisfied by the use of higher-tensile-strength resins or by increasing the standoff distance of the explosive charge. Another approach is the incorporation of z-direction, through-thickness fiber reinforcement, which serves to increase the through-thickness tensile strength of the FRP laminate. It is important to recognize that a z-direction reinforced laminate will mitigate delamination of plies within the laminate; however, adequate anchorage to the substrate must be established in order to prevent disbanding at the laminate substrate interface.

Research has shown that dynamic high-strain-rate explosive blast loading, as well as projectile impact, develops through-thickness stress wave propagation, which promotes resin matrix-dominated and fiber/resin interface-dominated failure modes in a composite laminate. The laminate experiences back-face delamination failure early in the event (be it ballistic impact or explosive blast loading) as a result of through-thickness stress wave reflections off the back face of the laminate.

C9.6.4 FRP-Strengthened, Reinforced Concrete Beams and Slabs. Damage expectations are provided in Table C9-6. The existing concrete substrate strength is an important parameter for bond-critical applications, including flexural or shear strengthening. It should possess the necessary strength to develop the design stresses of the FRP system through-bond. The substrate should exhibit sufficient direct tensile and shear strength to transfer load into the FRP system. The tensile strength of the concrete should be at least 200 psi (1.4 MPa) as determined by using a pull-off-type adhesion test as in ACI 503R-93 (ACI 1993, reapproved 2008) or ASTM D4541-09e1 (ASTM 2009). FRP systems should not be used where the concrete substrate exhibits a compressive strength, f'_c, less than 2,500 psi (17 MPa).

In order to overcome bonding problems associated with inadequate development of a chemical or adhesive bond between a reinforced concrete member (i.e., beam or slab) and an FRP laminate, or problems associated with the soundness and tensile strength of the concrete substrate, mechanical anchorage should be utilized whenever possible. Such mechanical anchorage should consist of rivets, mechanical fasteners with threaded inserts, or other mechanical shear ties to avoid (1) peeling, disbonding, or delamination of the FRP laminate from the concrete substrate at the laminate-to-concrete interface; or (2) shear and normal stress failure of the concrete substrate below the laminate-to-concrete interface. Mechanical anchorages can be effective in increasing load transfer between the reinforced concrete member and the FRP laminate. The performance of any mechanical anchorage system should be substantiated by testing when possible.

TABLE C9-7. QUALITATIVE DAMAGE EXPECTATIONS FOR FRP-STRENGTHENED, STEEL REINFORCED MASONRY WALLS

Mechanism	Superficial	Moderate	Heavy	Hazardous
Detachment at substrate	No	Limited	Yes	Yes
Detachment at FRP	No	Limited	Yes	Yes
FRP ply delamination	No	Limited	Yes	Yes
FRP ply edge delamination (peeling)	No	Limited	Yes	Yes
Fiber rupture of breakage	No	Limited	Yes	Yes
FRP lap splice failure	No	Limited	Yes	Yes
Matrix cracking	No	Limited	Yes	Yes
Matrix crazing	No	Limited	Yes	Yes

TABLE C9-8. QUALITATIVE DAMAGE EXPECTATIONS FOR FRP-STRENGTHENED, UNREINFORCED MASONRY WALLS

Mechanism	Superficial	Moderate	Heavy	Hazardous
Detachment at substrate	No	No	Yes	Yes
Detachment at FRP	No	No	Yes	No
FRP ply delamination	No	No	Yes	No
FRP ply edge delamination (peeling)	No	No	Yes	No
Fiber rupture of breakage	No	No	Yes	No
FRP lap splice failure	No	No	Yes	No
Matrix cracking	No	No	Yes	No
Matrix crazing	No	No	Yes	No

Alternatively, 90-deg anchorage zones can be provided, similar to what is performed in the case of U wraps. The extension of such zones can be defined by utilizing the procedure used for classical FRP applications, considering the tensile resistance of the substrate.

C9.6.5 FRP Strengthened Masonry Walls. Damage expectations are provided in Tables C9-7 and C9-8.

C9.6.6 FRP Concrete Column Confinement Requirements and Limitations. Reinforced concrete columns can be confined with FRP to enhance ductility and strength (shear and flexural). The design methods, and underlying mechanics, of ductility and strength enhancement are fundamentally different. Confinement designed to enhance strength will not necessarily provide appreciable, or predictable ductility improvement, and vice versa. This edition of the Standard only addresses ductility enhancement, which is likely to be the most important application of FRP confinement for blast-resistant design applications. More information can be found in Bank (2006).

Section 9.6.6 requires the confinement to be detailed to provide a ductility ratio of at least 6, which is a common minimum ductility for seismic FRP confinement. In keeping with the general approach of Chapter 9, the column should be designed such that there is a decreasing ductility demand as LOP increases. Thus, an LOP IV column with FRP confinement will be designed to be capable of achieving a ductility ratio of at least 6, even though the expectation is that the column will remain elastic under the blast load, providing a significant margin against uncertainty.

C9.6.7 FRP Solid Sections. In thin laminated plate bending theory, the contribution to out-of-plane plate deformations by transverse shearing stresses τ_{yz}, and τ_{xz} as well as the transverse normal stress, σ_{zz}, is negligibly small. Said another way, in thin laminated plate bending theory, applied transverse loads are reacted by in-plane shear and normal stresses, such that the out-of-plane transverse stresses τ_{yz}, τ_{xz} and σ_{zz} are negligible.

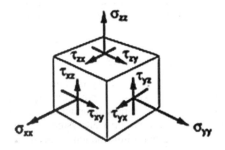

However, as a laminated plate becomes thicker, the transverse shear stiffness increases in relation to the axial and bending stiffnesses. Consequently, in "thick" laminates, the magnitude of transverse shearing stresses may no longer be considered negligible in reacting out-of-plane applied loads. Thus, a "thick" laminate is one in which the magnitude of transverse stresses (i.e., τ_{yz}, τ_{xz}, and σ_{zz}) is no longer negligible in relation to the magnitude of all in-plane stress components to the extent that the transverse stress components become significant terms in any 3-D (six stress component) laminate failure criteria.

For fiber-dominated failure modes in an advanced FRP composite material (as identified in Table C9-5), a maximum strain criterion may be utilized as defined below. The nondimensional factor, η, identifies the extent to which any ply within an FRP composite architecture exhibits first-ply failure. A value of $\eta < 1.0$ signifies that no ply in the laminate has failed. A value of $\eta = 1.0$ signifies that the laminate has experienced first-ply failure, whereas a value of $\eta > 1.0$ signifies failure of more than one ply within the FRP laminate.

$$\eta\varepsilon_{11}^{CU} \leq \varepsilon_{11} \leq \eta\varepsilon_{11}^{TU}$$
$$\eta\varepsilon_{22}^{CU} \leq \varepsilon_{22} \leq \eta\varepsilon_{11}^{TU}$$
$$|\varepsilon_{12}| \leq \eta\varepsilon_{12}^{SU}$$

For each loading condition, the strains in each ply are compared to these criteria. The failure mode and the first ply to fail under the applied loading condition are identified by whichever strain reaches its limiting value first. The limiting strains, $\eta\varepsilon_{11}^{cu}$, $\eta\varepsilon_{11}^{tu}$, $\eta\varepsilon_{22}^{cu}$, $\eta\varepsilon_{22}^{tu}$, and $\eta\varepsilon_{12}^{su}$ (where the superscripts cu, tu, and su represent compressive ultimate, tensile ultimate, and shear ultimate strains, respectively) are the specified maximum strains to be permitted in any ply. Such advanced FRP composite material properties (without the η factor) are contained in *MIL Handbook 17*, Vol. 2 (ASTM 2002b). When the material properties associated with an advanced FRP composite architecture are not available in *MIL Handbook 17*, Vol. 2, then such information may be obtained by testing a unidirectional laminate in uniaxial loading in accordance with the statistical sampling requirements identified in *MIL Handbook 17*, Vol. 1 (ASTM 2002a). For example, in the case of axial strain, ε_{11}, a statistically based unidirectional ultimate strain may be used, as in *MIL Handbook 17*, Vol. 2.

However, if a more detailed evaluation is necessary or if the failure mode of an advanced FRP material is matrix- or interface-dominated (as identified in Table C9-5), rather than fiber-dominated, then a "separate mode" failure criteria for determining first-ply failure, such as Hashin (*MIL Handbook 17*, Vol. 3, Section 5.4.1; ASTM 2002c), is warranted. More advanced failure criteria for FRP construction are contained in Hinton et al. (2004). Other specifications and standards for FRP construction are contained in Traceski (1990).

C9.6.7.1 Section 9.6.7.1 identifies the FRP element damage levels associated with each level of protection (LOP) as discussed in Section 3.3. The nondimensional factor, η, varies from 0.5 to 1.0 and identifies the extent to which any ply within an FRP composite architecture may approach first-ply failure. A value of $\eta = 1.0$ signifies that the FRP construction is allowed to reach first-ply failure.

C9.7 OTHER MATERIALS

C9.7.1 Aluminum. The majority of aluminum structures are comprised of custom extrusions, precluding the use of standard details for internal joinery and support anchorage. For aluminum punched windows and curtain wall systems, Chapter 8 and its Commentary give general information on detailing.

C9.7.3 Cold-Formed Steel Framing. Additional information on blast-resistant design of cold-formed steel wall panels can be found in UFC 3-340-02, Section 5–34: Special Considerations, Cold-Formed Steel Panels (DoD 2008).

Additional information on exterior panels can be found in DiPaolo and Woodson (2006).

CONSENSUS STANDARDS AND OTHER REFERENCED DOCUMENTS

American Concrete Institute (ACI). (2008). Building Code Requirements for Structural Concrete (ACI 318-08) and Commentary (ACI 318R-08). ACI, Farmington Hills, Mich.

ACI. (1993). Use of Epoxy Compounds with Concrete, ACI 503R-93 (reapproved in 2008). ACI, Farmington Hills, Mich.

American Institute of Steel Construction (AISC). (2010). Specifications for Structural Steel Buildings, ANSI/AISC 360-10. AISC, Chicago, Ill.

ASTM International (ASTM). (2009). Standard Test Method for Pull-Off Strength of Coatings Using Portable Adhesion Testers, ASTM D4541-09e1. ASTM, West Conshohocken, Pa.

ASTM. (2009). Standard Specification for Deformed and Plain Carbon Steel Bars for Concrete Reinforcement, ASTM A615-09b/A615M-09b. ASTM, West Conshohocken, Pa.

ASTM. (2006). Standard Specification for Low-Alloy Steel Deformed and Plain Bars for Concrete Reinforcement, A706/A706M-06a. ASTM, West Conshohocken, Pa.

ASTM. (2002a). *The Composite Materials Handbook MIL 17* (MIL-HDBK-17-3F), Vol. 1, Polymer Matrix Composites: Guidelines for Characterization of Structural Materials. ASTM, West Conshohocken, Pa.

ASTM. (2002b). *The Composite Materials Handbook MIL 17* (MIL-HDBK-17-3F), Vol. 2, Polymer Matrix Composites: Material Properties. ASTM, West Conshohocken, Pa.

ASTM. (2002c). *The Composite Materials Handbook MIL 17* (MIL-HDBK-17-3F), Vol. 3, Polymer Matrix Composites: Materials, Usage, Design and Analysis. ASTM, West Conshohocken, Pa.

Bank, L. C. (2006). *Composites for construction: Structural design with FRP materials*. John Wiley & Sons, Inc., New York.

Bank, L. C., Gentry, T. R., Thompson, B. P., and Russell, J. S. (2003). "A model specification for FRP composites for civil engineering structures." *J Const. and Building Mat.*, 17, 405–437.

DiPaolo, B. P., and Woodson, S. C., "An overview of research at ERDC on steel stud exterior wall systems subjected to severe blast loading." *Proc., ASCE/SEI Structures Congress, 2006.* ASCE, Reston, Va.

Dove, A. B. (1983). "Some observations on the physical properties of wire for plain and deformed welded wire fabric." ACI Journal Technical Paper Title No. 80-41. ACI, Farmington Hills, Mich.

Flathau, W. J. (1971). *Dynamic test of large reinforcing bar splices.* U.S. Army Corps of Engineers Waterways Experiment Station, Vicksburg, Miss.

Hinton, M. J., Kaddour, A. S., and Soden, P. D. (Eds.). (2004). *Failure criteria in fibre reinforced polymer composites: The world-wide failure exercise.* Elsevier Science, Amsterdam.

International Code Council (ICC). (2006). International Building Code, ICC, Washington, D.C.

Sierakowski, R. L., and Chaturvedi, S. K. (1997). *Dynamic loading and characterization of fiber-reinforced composites.* John Wiley & Sons, Inc., New York.

Traceski, F. T. (1990). *Specifications & standards for plastics & composites.* American Society for Metals International, Metals Park, Ohio.

U.S. Army Corps of Engineers (USACE). (2006). Single-Degree-of-Freedom Blast Effects Design Spreadsheets (SBEDS) Methodology Manual, PDC-TR 06-01. Protective Design Center, USACE, Omaha, Neb.

USACE. (2006) User's Guide for the Single-Degree-of-Freedom Blast Effects Design Spreadsheets (SBEDS), PDC-TR 06-02. Protective Design Center, USACE, Omaha, Neb.

U.S. Department of Defense (DoD). (2008). Structures to Resist the Effects of Accidental Explosions, UFC 3-340-02, <http://www.wbdg.org/ccb/DOD/UFC/ufc_3_340_02_pdf.pdf> [May 12, 2011].

DoD. (2002). Design and Analysis of Hardened Structures to Conventional Weapons Effects, UFC 3-340-01 (for official use only).

U.S. Department of Energy (DoE). (1992). A Manual for the Prediction of Blast and Fragment Loadings on Structures, DOE/TIC-11268 (unclassified, limited distribution), <www.osti.gov/bridge/product.biblio.jsp?osti_id=5892901> (May 12, 2011].

Weerth, E. (1989). MTL Technical Report 89-23, Composite Infantry Fighting Vehicle Program. U.S. Army Materials Technology Laboratory, Watertown, Mass.

Wire Reinforcement Institute (WRI). (2006a). *Manual of standard practice—Structural welded reinforcement,* WWR-500. WRI, Hartford, Conn.

WRI. (2006b). *Structural detailing manual,* WWR-600. WRI, Hartford, Conn.

Zehrt, W. H., and Lahoud, P. M. (1994). "Acceptable reinforcing steel splices for blast resistant concrete structures designed in accordance with TM 5-1300." *Minutes of the Twenty-Sixth Department of Defense Explosives Safety Seminar, Miami, FL, August 16–18, 1994.*

Zureick, A.-H., Bennett, R. M., and Ellingwood, B. R. (2006). "Statistical characterization of fiber-reinforced polymer composite material properties for structural design." *ASCE J. Struct. Eng.,* 132(8), 1320–1327.

Chapter C10
PERFORMANCE QUALIFICATION

C10.1 SCOPE

This chapter of the Standard provides procedures and standards for the qualification of performance of security-related components and devices. Reference is made to consensus standards (e.g., ASTM) and documents prepared by others that are available at the time of publication. The minimum performance levels, standoff distances, and explosive weights presented in these standards and documents vary and the user should adjust these variables to suit her/his needs.

The analytical and testing methods presented in this chapter are subject to the approval of the Authority Having Jurisdiction. The responsible design professional should seek the approval of the proposed method(s) of performance qualification in advance of numerical or physical simulations.

Tests and analyses performed to satisfy the requirements of this chapter should be designed with due consideration of the anticipated longevity, environmental effects, temperature effects, and other factors that are likely to affect the performance of materials and structural systems over their anticipated service lives.

C10.2 PEER REVIEW

It might be appropriate to conduct a third-party review of a blast design or component/equipment qualification undertaken in accordance with the procedures set forth in this Standard. The decision to require peer review shall be that of the Authority Having Jurisdiction. The format of the peer review shall also be decided by the Authority Having Jurisdiction.

The need for peer review might be triggered by a facility with unique structural features, a facility of high importance or economic value, or by a design team with minimal experience in blast design.

The peer review team should be independent of the design team to add objectivity to the process and value to the final product. For review of a blast design, the team should be formed prior to the schematic design phase and the development of building-specific design criteria and the choice of structural framing systems.

C10.3 SITE PERIMETER COMPONENTS

C10.3.1 Performance Qualification by Full-Scale Testing.
Sufficient numerical and video data must be collected to completely characterize the response of the test article and the impacting vehicle to enable calculation of the energy absorbed by the soil and foundation, the impacting vehicle, and the test article. Accelerometers, displacement transducers, and strain gages should be used to record numerical data. High-speed digital cameras should be used to record visual data.

The impact conditions shall be reported using the measured speed, computed equivalent speed, and angle of attack of the test vehicle. The measured speed is that of the test vehicle at impact. The computed equivalent speed shall be based on the energy absorbed by the test article and the mass of the test vehicle. The performance of the site perimeter component shall be reported in terms of damage to the test article, test article transient and permanent deflection, postevent functionality of the test article, and test vehicle penetration.

Consideration should be given to the development of a large-deformation mathematical model of the test article and its supporting foundation (including soil) and the impacting vehicle to validate nonlinear numerical models and procedures and to enable extrapolation of the test results to similar site-perimeter components and loading environments.

Detailed procedures for determining the crash performance of perimeter barriers and gates are presented in ASTM F2656-07 (ASTM 2007).

C10.3.2 Performance Qualification by Analysis and Design.
The performance of site-perimeter components for vehicular attack can be qualified by analysis and design using nonlinear, large-deformation finite element analysis computer codes, hydrocodes, or other validated numerical analysis methods. The computational procedures and constitutive models used in the analysis must have been validated by prior analysis of similar structures subjected to similar loading environments. The numerical model used for performance qualification must include:

- Site-perimeter structure
- Foundation and subsurface materials
- Impacting vehicle.

The report shall include but is not limited to the information listed in the Standard. The studies used to validate the computational procedures and the constitutive models should be provided as an appendix to the report.

C10.4 Building Structural Components

C10.4.1 Performance Qualification by Full-Scale Testing.
Structural components such as proprietary beam-column connections can be qualified by arena or shock tube testing. Sufficient numerical and video data must be collected to completely characterize the charge and the response of the component. Pressure transducers shall be used to monitor the airblast loading on the component. Accelerometers, displacement transducers, and strain gages should be used to record the response of the test article. High-speed digital cameras should be used to record visual data.

A sufficient number of tests should be performed to provide reliable assessments of performance characteristics. The number of tests will be a function of the anticipated dispersion in the response and the required confidence in the results.

The blast loading conditions shall be reported using the charge size, standoff distance, elevation, and orientation. The performance of the structural component shall be reported in terms of damage to the component, component transient and permanent deflection, and penetration of the component by fragments and debris.

Consideration should be given to the development of a large-deformation mathematical model of the component to validate nonlinear numerical models and procedures and to enable extrapolation of the test results to similar structural components and loading environments.

Work should be undertaken in laboratories or on ranges with a proven capability to perform such tests.

C10.4.2 Performance Qualification by Analysis and Design. The performance of a structural component under airblast and debris loadings can be qualified by analysis and design using nonlinear, large-deformation finite element analysis computer codes, hydrocodes, or other validated numerical analysis methods. The computational procedures and constitutive models used in the analysis must have been validated by prior analysis of similar structures subjected to similar loadings.

The report shall include but is not limited to the information listed in the Standard. The studies used to validate the computational procedures and the constitutive models should be provided as an appendix to the report.

C10.5 SHIELDING STRUCTURES

C10.5.1 Performance Qualification by Full-Scale Testing. Sufficient numerical and video data must be collected to completely characterize the charge and the response of the test article. Pressure transducers shall be used to monitor the loading on both faces of the shielding structure so that the reduction in blast loading can be quantified. Accelerometers, displacement transducers, and strain gages should be used to record the response of the test article. High-speed digital cameras should be used to record visual data.

A sufficient number of tests should be performed to provide reliable assessments of performance characteristics. The number of tests will be a function of the anticipated dispersion in the response and the required confidence in the results.

The blast loading conditions shall be reported using the charge size, standoff distance, elevation, and orientation. The performance of the shielding component shall be reported in terms of damage to the test article, test article transient and permanent deflection, penetration of the test article by fragments and debris, and reduction in blast effects at the location of the test article (noting that the blast wave might re-form on the far side of the shielding structure).

C10.5.2 Performance Qualification by Analysis and Design. The performance of shielding structures under airblast and debris loadings can be qualified by analysis and design using nonlinear, large-deformation finite element analysis computer codes, hydrocodes, or other validated numerical analysis methods. The computational procedures and constitutive models used in the analysis must have been validated by prior analysis of similar structures subjected to similar loadings. The numerical model used for performance qualification must include:

- Shielding structure
- Foundation and subsurface materials
- Impacting vehicle.

The report shall include but is not limited to the information listed in the Standard. The studies used to validate the computa-

tional procedures and the constitutive models should be provided as an appendix to the report.

C10.6 BUILDING FAÇADE COMPONENTS

C10.6.1 Glazing and Glazing Systems. Detailed procedures for testing glazing components and glazing systems (including curtain walls) are presented in ASTM F1642-04 (2010) (ASTM 2010) for determining the performance of a glazing, glazing system, or glazing system retrofit subjected to airblast loading. Laminated glass can be qualified as blast-resistant glazing by analysis and design using an equivalent 3-sec duration face loading provided that (1) the glazing is adhered to its supporting frame using structural silicone or glazing tape, (2) the framing members that support the edges of the glazing are stiff, and (3) the fasteners that attach the glazing to the structural frame are stronger than the glazing, all per ASTM F2248-03 (ASTM 2003).

Alternate procedures are presented in Section 10.6.1.2 for qualifying glazing and glazing systems by analysis and design.

C10.6.2 Doors. General test and data recording and reporting procedures for blast testing of doors in open-air arenas can be adapted from those developed for glass and glazing systems in ASTM F1642-04 (2010) (ASTM 2010).

Metal doors can be qualified as blast-resistant by imposing static pressures on a full-size test specimen by (1) installing and sealing the test specimen into or against one face of a test chamber, (2) supplying air to the chamber at a ratio to maintain the required pressure differential across the specimen, (3) digital and video recording the deflections, deformations, and strains in the test specimen, and (4) documenting the damage to the components of the test specimen and test frame, all per ASTM F2247-11 (ASTM 2011).

Alternate procedures are presented in Section 10.6.2.2 for qualifying blast-resistant doors by analysis and design.

C10.7 BUILDING NONSTRUCTURAL COMPONENTS

C10.7.1 Performance Qualification by Full-Scale Testing. Nonstructural components and equipment can be qualified by arena or shock tube testing. Sufficient numerical and video data must be collected to completely characterize the charge and the response of the component. Pressure transducers shall be used to monitor the airblast loading on the component. Accelerometers, displacement transducers, and strain gages should be used to record the response of the test article. High-speed digital cameras should be used to record visual data.

The blast loading conditions shall be reported using the charge size, standoff distance, elevation, and orientation. The performance of the component shall be reported in terms of damage to the component, component transient and permanent deflection, and penetration of the component by fragments and debris.

Consideration should be given to the development of a large-deformation finite element, hydrocode, or other validated numerical analysis model of the component to validate nonlinear numerical models and procedures and to enable extrapolation of the test results to similar structural components and loading environments.

Work should be undertaken in laboratories or on ranges with a proven capability to perform such tests.

C10.7.2 Performance Qualification by Analysis and Design. The performance of a nonstructural component or piece of equipment under airblast and debris loadings can be qualified by analysis and design using nonlinear, large-deformation finite element analysis computer codes, hydrocodes, or other validated numerical analysis methods. The computational procedures and constitutive models used in the analysis must have been validated by prior analysis of similar structures subjected to similar loadings.

The report shall include but is not limited to the information listed in the Standard. The studies used to validate the computational procedures and the constitutive models should be provided as an appendix to the report.

REFERENCES

ASTM International (ASTM). (2011). Standard Test Method for Metal Doors Used in Blast Resistant Applications, ASTM F2247-11. ASTM, West Conshohocken, Pa.

ASTM. (2010). Standard Test Methods for Glazing and Glazing Systems Subject to Air Blast Loadings, ASTM F1642-04 (2010). ASTM, West Conshohocken, Pa.

ASTM. (2007). Standard Test Method for Vehicle Crash Testing of Perimeter Barriers and Gates, ASTM F2656-07. ASTM, West Conshohocken, Pa.

ASTM. (2003). Standard Practice for Specifying an Equivalent 3-Second Duration Design Loading for Blast Resistant Glazing Fabricated with Laminated Glass, ASTM F2248-03. ASTM, West Conshohocken, Pa.

INDEX

structural systems 23–27, 71–80; application of loads 25–26; axial load effects 24–25; concrete elements 24; concrete frame with concrete shear wall systems 26, 78; concrete moment frame systems 26, 78; connections and joints 25, 77; consensus standards 27; design 26; elements modeling 74–77; flexure 24, 74–75; forces in bending elements 75–76; instability 25, 76; masonry elements 24; mass 26; materials 74; modeling and analysis 23–26, 72–78; negative phase 77–78; precast tilt-up with concrete shear wall systems 26, 78–79; pressure-impulse charts 72–73; reinforced concrete elements 24; reinforced masonry bearing shear walls 26, 79; response characteristics 26–27, 79–80; shear 24; single element response analysis 73–74; steel braced frame systems 26, 78; steel moment frame systems 26, 78; strength increase 23–24, 74; structural design 78–79; structural steel elements 24

symbols and notation 3–4

threat reduction 5–6, 50–53

threats: accidental 5–6, 50–51; internal 29; malicious 5–6, 50, 51–52; standoff 51–53; vehicle barriers 53

vehicle barriers 6, 53

walls 37, 38; cast-in-place 33, 91; concrete 78–79, 97; FRP strengthened masonry 39; isolating from internal threats 29; masonry 33, 79, 91; non-load-bearing exterior 32–33, 91; precast 33, 78–79, 91; pretensioned and posttensioned concrete 33, 91; steel 33, 91; tilt-up 33, 78–79, 91. *see also* shear wall systems

windows 31–32, 85–95

wood 39

Date Due

BRODART, CO.   Cat. No. 23-233-003   Printed in U.S.A.